AFRICAN POLITICAL, ECONOMIC, AND SECURITY ISSUES

THE POLITICAL ECOLOGY OF SOIL EROSION IN WEST AFRICA

AN HISTORICAL AND CONTEMPORARY PERSPECTIVE

African Political, Economic, and Security Issues

Additional books in this series can be found on Nova's website under the Series tab.

Additional e-books in this series can be found on Nova's website under the e-book tab.

Environmental Research Advances

Additional books in this series can be found on Nova's website under the Series tab.

Additional e-books in this series can be found on Nova's website under the e-book tab.

AFRICAN POLITICAL, ECONOMIC, AND SECURITY ISSUES

THE POLITICAL ECOLOGY OF SOIL EROSION IN WEST AFRICA

AN HISTORICAL AND CONTEMPORARY PERSPECTIVE

MICHAEL O'NEAL CAMPBELL

nova publishers
New York

Library of Congress Cataloging-in-Publication Data

ISBN: 978-1-62417-258-8

Published by Nova Science Publishers, Inc. † New York

CONTENTS

PREFACE

This book is based on literature analyses and field research I conducted for my PhD at Wye College (now Imperial College at Wye), University of London. The objective was to adopt a broad contextual and case study approach, using some relevant theoretical strands that have developed over the last few decades, related to the meta-theory of West African nature-society relations and environmental change. Soil erosion is a serious problem for the development of West Africa. It has social, political and economic origins, and as such is not a purely natural phenomenon. As people are the basic problem, the human situation must be emphasised. Due to the varying socio-cultural contexts that exist in any region, generalisations are difficult to make. Hence, a contextual approach is important. Many paradigms developed in the past do not adequately explain the origin and causes of soil erosion. There is no single solution to the incidence of accelerated erosion. Political, economic, socio-cultural and natural contexts are vital areas of action. An open mind is important. Finally, in this, as in any problem, for any real and lasting solution the cooperation and commitment of all the people involved is essential.

INTRODUCTION

Soil erosion is one of the most important forms of environmental change in the modern world, and has extremely important historical, social, scientific and future implications, especially in Africa. For an effective assessment of the roots and solutions for soil erosion and related problems, multidisciplinary and interdisciplinary methods must be explored. This book examines the issue of soil erosion from political, economic, social and ecological perspectives, and through the lens of multidisciplinary frameworks such as political ecology, environmental sociology and geography. The objective is to present soil erosion as an inherently multifaceted issue, the better to derive relevant solutions to this vital global problem and to illustrate these issues in a well selected case study. Five key issues are presented and analysed: (1) political ecology as a tool of appraisal, analysis and evaluation, which assumes political power and social relations are factors for environmental change; (2) social structures such as the household, gender relations and methods of participation; (3) the natural factors for soil erosion; (4) the social factors for soil erosion; and (5) the extent to which a multidisciplinary approach can analyse a relevant case study in an African landscape. The dominant perspective emerging from this analysis results from the assessment of the commonly dominant social paradigm, which sees local people as degraders of the environment. It also assesses the opposed political economy approach that points to local people as helplessly marginalised and forced to degrade marginal environments, rather than as marginalised *and* innovative. The result is a more broadly based paradigm that takes from both sides of the divide, and seeks to chart a forward path.

The first chapter explores political ecology as a system of thought and a tool of analysis. As such, it examines the definition, history, background paradigms (e.g. political economy), coverage and assumptions of political ecology. There are three basic assumptions: (1) the unequal distribution of the costs and benefits of environmental resource access; (2) the possibility that such inequality may exacerbate the inequalities in the society; and (3) the resultant relations have strong political implications, which may influence social relations with the environment. The paradigmatic and literary development of political ecology from the 1970s includes its emergence from the structuralist perspectives in social science and the recent, broader definitions of ecology. These touch on the relation between political ecology and cultural ecology, geography, anthropology, political economy and environmental science. To complement the structuralist basis of political ecology and enable its firm social foundation, key complementary building blocks are structuration theory, the balance between structural and actor oriented approaches to social analysis, and applications to environmental relations.

The second chapter applies these ideas to actual structures and actor based scenarios in the rural environmental context. These are livelihood systems, inter- and intra- household relations and gender issues. This leads to a discussion of the multidisciplinary paradigm of sustainable development, which is a argued to be a concept, a political objective and a principle for policy action. It is examined from the viewpoints of economists, ecologists and "green politics". The economists' is the maximisation of the net benefits of economic development, subject to the maintenance of services and the quality of resources over time, with some advocating social equality, environmental sensitivity, and a reduction of the maximum growth objective. Ecologists adhere to a definition that emphasises the support or enhancement of biological diversity and productivity, promotion of stronger resilience of ecosystems to externalities and natural, methods for agriculture. The green position is of reduced economic growth (or even "zero growth"), the better to halt environmental degradation.

These arguments provide the basis of the third chapter, which examines the political ecology of the relationship between agricultural workers and political actors, and the relation of these to agricultural development. The key issues in this chapter are how local people, as actors, react to structures that may enable and/or constrain them in the creation of other structures, such as households and livelihoods. Different degrees of social participation in development are examined, listed in order of greater strength as manipulative, passive, consultative, managerial, incentive based, functional, interactive participation, and self-mobilisation. This links to a discussion of a small farmer approach to agricultural development policy (termed agrarian populism), Indigenous Technical Knowledge (ITK) (or cultural knowledge or Rural People's Knowledge RPK) and the merits of participation. What follows is an analysis of the importance of social participation to the African savanna, manifested by village level committees, land disputes and other conflicts, and agricultural projects.

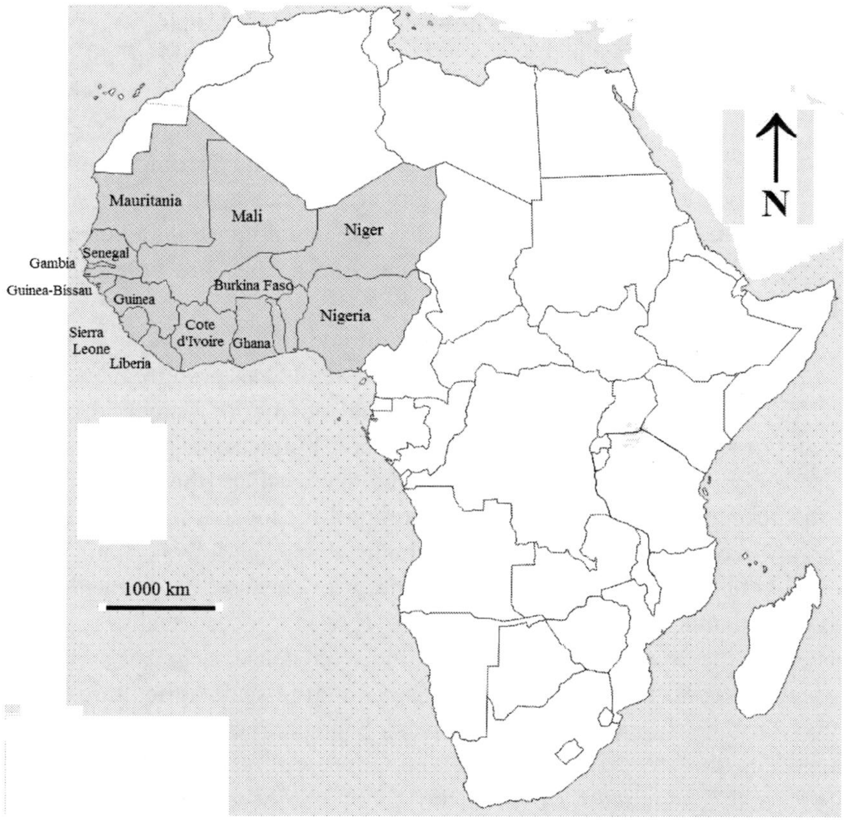

Figure 1. The Region of West Africa within Africa (United Nations Classification).

Chapter 4 opens the topic of soil erosion, firstly as a natural phenomenon, looking at definitions, processes and incidence, and then as a social issue, related to externalities and contributory factors, and accelerated erosion in the case of West Africa (Figure 1). Soil erosion connects to population increase, agricultural intensification, famine and the Malthusian hypothesis of limited resources. Also examined are the natural factors for soil erosion (geology, relief, climate, soils, and vegetation), the main soil types of West Africa and the Universal Soil Loss and Wind Erosion Equations. The natural onsite and offsite effects of soil erosion, including desertification, and the social factors of soil erosion are then discussed. The rest of the chapter looks at accelerated soil erosion in West Africa. The land use patterns (traditional agriculture, i.e. shifting cultivation, sedentary rainfed agriculture and cattle herding) result from deep political, economic and cultural structures, and relate to population growth, agricultural intensification, forestry and urbanisation.

Chapter five takes Ghana as a case study, examining the traditional and modern agricultural systems of the country and other factors for soil erosion. It also looks at political decentralisation, development organisations and gender issues as important factors in the efforts to combat deforestation and consequent soil erosion. For example, poverty forces poor people to exploit, and adapt to marginal lands and use fuelwood, denuding rural and urban areas. Logging companies have incentives to cut tree stands, but political powers frequently ignore their activities.

Chapter six takes a case study of a field and image based analysis of the coastal savanna of Ghana. In this area, flat land was preferred by all land users, but its limited availability, the unequal powers of stakeholders (small rainfed farmers versus large scale farmers and builders, men versus women, old land users versus young land users), the dry weather (the driest in Ghana) and the strongly seasonal, convectional rainfall meant that many steep slopes were utilised by the weaker, marginalised stakeholders. Soil erosion, waterlogging and desiccation resulted, despite attempts at innovation. The study found that there was substantial deforestation, especially in lowland areas, and the average slope and altitude of dense vegetation increased between 1960 and 1996. Building landuse and large-scale agriculture replaced rainfed farming in many lowland areas; hence, most of the deforestation occurred on the steeper slopes of marginalised cultivators. The soils of the study area are examined for their fertility, vulnerability to erosion, desiccation and waterlogging, and the scientific soil classification is compared and contrasted with local farmer knowledge. The problems of farm clearance increased social, economic and environmental costs, and further exacerbated the problems faced by marginalised farmers.

The book concludes that the political ecology methodology is an incisive analytical tool, provided it is sufficiently flexible to incorporate both contextual variables and scaled relations at actor, social structure, landscape and government levels. Important questions emerge concerning the role of a political ecology approach in actually contributing to the solution of the problems of social relations and soil erosion. This will hinge on the representation of the different facets of the problem within the theoretical framework, and the extent to which relations of the real world are documented and analysed.

List of Figures

Chapter 1

Political Ecology, Its Roots and Complements

1.0. Introduction

This chapter examines political ecology, structuration theory, structuralism and actor oriented paradigms, as a background of the study of socio-environmental relations that contribute to or prevent soil erosion. Topics discussed include the history of these subjects, their similarities, differences and complementary roles and applications. There is particular emphasis on those topics that are relevant to livelihoods, households, gender, and socio-environmental relations, the topics of the next chapter. These topics are relevant to soil erosion, because the political, economic and socio-cultural relations, manifested by power, access and networking, determine the agricultural possibilities, and the intensification of activities in marginal, fragile or agriculturally optimal lands.

1.1. Political Ecology

The term "political ecology" was first mentioned by Frank Thone in 1935 (Nature Rambling: We Fight for Grass, The Science Newsletter 27 717, Jan. 5:14). The anthropologist Eric R. Wolf reintroduced the term in 1972 (Wolf 1972). There is no single agreed definition of political ecology (Hanna et al. 2008), but most definitions relate to social relations, group-related power differences, knowledge, conflicting resource access and nature-society relations. Political ecology may be "an approach to, but far from a coherent theory of, the complex metabolism between nature and society" (Watts 2009: 545). Another definition is that it "seeks to understand the complex relations between nature and society through a careful analysis of what one might call the forms of access and control over resources and their implications for environmental health and sustainable livelihoods" (Watts 2000: 257). Another author defines political ecology as an "overarching frame of inquiry for exploring the politics of natural resource access and use at multiple levels over time" (Wilshusen (2003: 41). Also, "political ecology examines the political dynamics surrounding material and discursive struggles over the environment in the third world" (Bryant (1998: 79). Hanna et al. (2008: 201) note that this definition, "by referencing both material and discursive struggles, highlights *two branches* of political ecology." These are the structural, which is focussed on the interelationship between "ecology and a broadly defined political economy" (as noted by Blaikie and Brookfield, 1987: 17), and the "poststructural, which focuses on the control over discourse, knowledge, and ideas" (also mentioned by Blaikie, 1999 and Watts, 2000) (Hanna et al. 2008: 201).

Whichever definition is used, political ecology as an academic discipline gives a broad perspective that integrates ecological sciences with political economy (Peet and Watts 1996), applied to issues such as: links between environmental degradation and socio-economic marginalisation; political conflict integrated with social and environmental issues; environmental conservation; socio-economic power relations; and environmental and social movements (Robbins, 2004; Watts 2009). Currently, political ecology has diverged into several strands including: environmental history (Grove 1995); science studies (Demeritt, 1998); actor-network theory (Braun and Castree 1998); gender theory (Agrawal 1998); discourse analysis (Escobar 1995); and Marxism (Leff 1995; O'Connor 1998).

Political ecology supplants its predecessor, cultural ecology, largely due to the failure of the latter discipline to prioritize "the broader political and economic forces impacting on local human-environment interactions" (Hanna et al. 2008: 201). The main difference between political ecology and cultural ecology was thus that political ecology focussed the direct interaction between social power and the environment, and also examined those socio-political issues surrounding, enabling and constraining the political-environmental interaction (Walker 2005; Hanna et al. 2008). Political ecology developed in the 1970s as an integration of political economy and ecology, combining the relation between natural resources and society, and relations between structures, groups and classes within society (Blaikie and Brookfield 1987). In the study of the relations between structures in society, political ecology was highly affected by neo-Marxist ideas, these emphasising the impact of the global capitalist system on production systems and class relations, and the relation of these with the environment. Neo-Marxist political ecology attempted not only to explain, but also to solve problems such as poverty in the Global South. A "structural political ecology" examined political, economic, social and environmental resource issues from both historical and scalar perspectives, from local to global (Hanna et al. 2008). Particularly important social issues concerned the localisation of power relations, these influenced by "the presence of markets, deep social inequalities, enduring conflict, and forms of cultural disintegration associated with their integration into a modern world system" (Watts 2000: 261).

Political ecology as initially crafted was increasingly seen as "narrow and structural, and thereby incapable of understanding complex political, economic or cultural processes" (Bryant and Goodman 2004: 347). Consequently, the format of political ecology began to change in the 1980s, from the overly structural or structural-determinist approach of neo-Marxism, which rigidly characterised the poor as invariably oppressed, within structured views of the environment and socio-environmental relations, to a theoretically flexible framework (especially post-structuralism and discourse theory) emphasising the different relations between different sets of actors and the multiple occurrences between these relations and the environment (Escobar 1994, 1996; Peet and Watts 1996; Bryant and Bailey 1997). Eventually, a post-Marxist, post structuralist political ecology emerged that prioritised broader cultural and political issues, to the detriment of rigid theoretical structures (Escobar 1996; Peet and Watts 1996b; Rocheleau et al. 1996; Bryant and Bailey 1997; Neumann 1998).

The post-structuralist view was centred on the importance of *discourse*, where discourse was defined as "a particular area of language use related to a certain set of institutions and expressing a particular standpoint" (Peet and Watts 1993: 228), or "a shared meaning of a phenomenon" (Adger et al. 2001: 683). Translated from the language of theory to practical application, discourses or understandings (for example of sustainable development) may differ and translate into the promotion of different actions to the neglect of others, favouring the powerful over the weak, with environmental relevance (Escobar 1996; Hanna et al. 2008). Included in the framework of multiple, competing or conflicting discourses is the concept of the plurality of knowledge, including that of the environment, and the idea that these knowledges are debatable (unlike the earlier, structuralist versions of political ecology), to the point that the environment itself is a social construction (Hanna et al. 2008). These subjective discourses are created and maintained by various actors (scientists, policy makers, local actors, economists) and both silence and promote related knowledges, construct perceptions about the environment, and shape policy and political ecology (Fischer 2000; Hanna et al. 2008).

According to Bryant and Bailey (1997), there are three assumptions that give political ecology relevance. Firstly, there is an unequal distribution of the costs and benefits linked to environmental change, due to socio-economic and political differences and inequalities. Secondly, the existing inequalities and differences are exacerbated or ameliorated, by the unequal distribution of costs and benefits: hence, change in the environment affects the political and economic system. Thirdly, both the unequal distribution of costs and benefits and the related change in pre-existing inequalities create externalities manifested by modified power relations. Developing from these dynamics, political ecology may provide an imperative or directive for more effective or balanced policy action: hence it may possess a "normative understanding that there are very likely better, less coercive, less exploitative and more sustainable ways of doing things" (Robbins 2004: 12). These contribute to superior environmental governance, greater understanding of the environmental decision making of communities within contexts of differing socio-economic and political structures, and further understanding of the role of inequality in policy, society and environmental relations.

1.2. Political Ecology, Geography, Anthropology and Conservation

These developments are vital, because an important justification of the development of political ecology concerns its movement beyond perceived flaws in related, older disciplines, the better to explain socio-environmental dynamics. A case in point is the overly structuralist nature of early political economy (Perry 2003: Ritzer 2008) and anthropology (Wolf 1997). Also, methodologies in cultural ecology, while inclusive of ecological variables and ethnographic detail, where often negligent of interactions between socio-economic and environmental variables (Perry 2003). Political ecology, as developed by geographers and anthropologists (Wolf 1972; Blaikie 1985,

Blackie and Brookfield 1987; Greenberg & Park 1994; Robbins 2004) used techniques explored by these disciplines (political economy, cultural ecology, anthropology) and moved further by developing a methodology for examining the interactions of environmental, political and socio-cultural variables. Simultaneously, it avoids one feature of these overly structuralist predecessors, namely the overt separation of variables into distinct disciplines and structures (Robbins 2004; Biersack 2006: Elwood 2010).

Through acknowledgment of its paradigmatic utility, political ecology has been strongly applied to diverse topics within the conservation sciences, topics including social political movements, natural resource conservation, local resource use politics, socio-economic marginalization and environmental degradation, consumption and production, environmental conflicts and identities, conservation and control of natural resources, the disempowerment of local stakeholders, the effects of limited state capacity, informal claims to resource use, and ambiguities of property rights (McCarthy 2002, 2005, 2006; Walker 2005, 2006; Robbins 2004; Pomeroy 2012). Concerning conservation, particularly important themes are the re-investigation of complex environmental issues, which led geographers towards "conceptualizing and articulating the relationships between structural processes and local contexts, and clearly establishing which methods and data may be best used to get at which aspects of these relationships" (Elwood 2010: 104). Implicit in this contribution was the articulation and description of new principles for action (Hanna, et al. 2008) and involvement of local people, who "have important environmental knowledge which could contribute to conservation" (Dove and Carpenter 2008: 4). In the application of political ecology, some authors have nevertheless observed that political ecology still pays insufficient attention to individual actors, a problem solved if structuration theory is included (Zimmerer 1996; Pomeroy 2012). The next section will therefore examine the concepts of actor and structure, and how these underpin the theory of structuration and complement political ecology.

1.3. Social Structures and Actors

This section looks at definitions, meanings and utility of social structures and structuralist methodologies, before examining actor oriented theories and then structuration theory. Structures are defined as any recurring pattern of social behaviour and the ordered interrelationships between the elements of a social system or society; examples being kinship, religious, economic, political and other institutions as well as associated norms, values and social roles (Marshall 1994). Hence, all societies may be dominated by structures (Durkheim 1982; Craib 1992). Giddens (1979) argues that structures exist in the minds of social actors, just as language exists in the minds of language users. Structures are also rules and resources which are embedded in systems of social interaction (Giddens 1984). The resources, allocative and authoritative, are supportive of the rules (Giddens 1979, 1984). Allocative resources are the material features of the environment and the means of production and produced goods. Authoritative resources comprise the organisation of social time-space, the production and reproduction of the body, and life chances (Craib 1992).

Pepper (1996) argues that any theory is structuralist if it holds that there are deep, unobservable, only subconsciously apprehended realities that may give rise to observable issues; these portray socio-economic reality as a superstructure of social beliefs, ideas, relations, institutions, practices and rituals, which quite rigidly determine the underlying economic base. Structural functionalism, the main sociological theoretical framework before the late 1960s, held that structures within society have different functions, and societal development causes ever greater structural and hence functional, complexity. Simple societies have few structures, with few functions, such as the kinship system; in the modern society, the family is just one specialist system among others, such as the economic, education and political systems (Craib 1992). Some social analysts over-emphasised the determining role of structures over actors. Craib (1992) notes that at the peak of structuralism, people were examined as the products of structures, rather than as free decision makers. This ignores how humans individually and socially, constantly change their own nature (as well as external nature) through history (Pepper 1996).

Other social theorists emphasise the actor oriented approach, largely as a post-structural deviation from emphasis on large scale social and historical processes in favour of a micro-sociology of subjective life, and associated linguistic and cultural processes (Benton and Redclift 1994). Long (1992: 5) suggests that the essence of an actor approach includes: the examination of the everyday life experiences and understandings of men and women, regardless of societal station; the placement of individuals at "centre stage" of multiple realities; variable practices; interfacial situations; and the knowledge and power of actors (Long 1992: 5). People are seen as active agents, rather than as disembodied social categories, classes, passive recipients or parts of structural logic (Long 1992). This approach enables an explanation of variable individual behaviours in similar scenarios. Nevertheless, a major criticism of actor oriented frameworks concerned the the excessive prioritisation of voluntarism. Some (mainly anthropological) studies exceeded bounds by neglecting the effect of structural influences on human behaviour and adopted an extreme form of methodological individualism that visualised social behaviour primarily in terms of individual motivations, intentions and interest (Long 1992). Additionally, related studies were criticised for being ethnocentric models of western social behaviour, based on the non-contextual individualism of utilitarianism (Long 1992).

The integrated analysis of structuration theory is intended to supersede these problems by approaching structures and individual agency as two ways of looking at the same phenomenon (Giddens 1979). This is a point identified by Redclift and Woodgate (1994), and by Benton and Redclift (1994:19) who note that structuration is "intended to incorporate and go beyond the structure/action polarity."

Basically instead of seeing structures and actors as separate entities (a dualism), they are perceived aspects of the same entity (a duality): when one looks at the duality one way, actors are seen, when one looks at the same duality from another angle, structures are seen. This is defined by Giddens as the duality of structure. Structuration theory postulates that social structures are created and reinforced by individual, decision making actors and these actors are also constrained and enabled by the structures of their own creation and those created by others.

This is relevant to political ecology, as political ecology does not focus enough on the decision making of individual actors. When combined with structuration theory it compensates for this omission (Zimmerer 1996). In terms of socio-environmental relations and agricultural development, farmers and other actors within and external to the agricultural system, reinforce or challenge structures and thereby change the conditions that enable and constrain socio-environmental relationships (Hakeem and Vemuri 1997; Rowe 1997; Woodgate and Redclift 1998).

Indeed, the environment may be evaluated as similar to a social structure that enables and constrains actor behaviour, and is modified by them. Benton and Redclift (1994:19) argue that structuration "although not invented to deal with environmental issues...can be readily applied here too because the environment itself can be thought of as a structure which both constrains and enables action *and* may be changed by social action," a point also stressed by Redclift and Woodgate (1994) and Woodgate and Redclift (1998).

1.4. Links between Social Structures, Actors and the Environment

Social structures that are particularly relevant to socio-environmental interfaces in the Global South are the household, socio-cultural norms (especially gender relations), the state/farmer relationship and livelihood activity patterns. These comprise relevant relationships between actors in creating, recreating, maintaining, and reacting to structures. They include hierarchical structures of decision making, which may promote and marginalise individuals, and develop and affect actors with variable abilities in coping with socio-economic differentiation (Scoones and Thompson 1994; Woodgate 1991, 1992; Redclift and Woodgate 1997; Woodgate and Redclift 1998). Solbrig (1993) exemplifies this position through the work of Westoby et al. (1989) who argue that people interacting with the environment for sustenance must make an inventory of the possible alternative states of the structure and of the possible transitions from one state to another. The example cited is of the environmental knowledges and livelihoods of the nomadic pastoralists of the African savannas. Other examples are of farmers managing shifting cultivation systems and broadening livelihood options, working with perceived resources and the transitions between them (for example fallow periods and soil fertility change, or the dynamics of alternative activity change).

This broad approach, termed holistic by Eger et al. (1996) or "holism" by Rowe (1997: 149) emphasises the importance of individual people as shapers of larger structures, and the role of understanding as an influence behind such activities. Also recognised is the socially grouped nature of knowledge, context and history. "Conceptual pluralism", defined as variations of knowledge, viewpoints and understanding, is also important (Norgaard 1994: 93). This encourages bio-cultural diversity, decentralisation and participation, which are important development issues (Chopra 1996; Eden and Parry 1996). The merit of the holistic approach lies in its ability to analyse chaotic relations in socio-environmental systems and structures without falling into an unnecessarily narrow and rigid viewpoint or methodology. If the objective is economic development, actor, power differentiation, structural change and intensive economic linkages may affect socio-environmental linkages for greater production (Solbrig 1993; Solbrig and Young 1993; Hatzius 1996a; Reenberg and Paarup Laursen 1997; Woodgate and Redclift 1998). Also important are the wider national and international policies and economic linkages that provide credit, subsidies, exchange rates and markets. Reflecting on these points, it is prudent that no single set of general management prescriptions are optimal; socio-environmental relations and projects are social constructions undergoing continual development (Solbrig 1993; Maass (1995).

The holistic, coevolutionary framework underpinning political ecology also has important implications for policy making (Woodgate 1992; Mertens (1994). In this practical application to empirical contexts for development policy and planning, Woodgate (1991: 180) suggests that two approaches are useful. The first is the introduction of new agricultural techniques accompanied by the appropriate knowledge and organisational structures necessary to sustain and react to the environment's responses to the new technology. The second is the encouagrement of new techniques from local knowledge and experience, pre-adapted to existing organisational structures. This, as noted by Redclift and Goodman (1991: 20) in reference to Woodgate's thesis of coevolutionary development, is the only development that can succeed in combating environmental degradation, while preserving the cultures - themselves dynamic and "evolutionary"- in which local knowledge is encoded. This Woodgate (1991: 181) terms the "sustainability dynamic." Policy makers should emphasise the relationships that contribute to changes in socio-environmental relations, by defining and promoting the quality of life of actors, at different spatial scales of relevance (for example regional, national and international) (Mertens 1994).

CONCLUSION

The broad, theoretical issues discussed in this chapter provide a basis for more detailed social analyses in the succeeding chapters. The key issue emerging from this discussion is the importance of a balance between social structures, individual actors, environmental features and interpretative methodologies. How do we assess these relationships, objects and subjects fairly and accurately? The answer to this question is to seek a deeper understanding of these features, the better to explain, suggest, apply and solve problems. The succeeding chapters tackle these issues.

Chapter 2

THE SOCIAL ANALYSIS OF RURAL DEVELOPMENT: LIVELIHOODS, HOUSEHOLDS AND GENDER RELATIONS

2.0. INTRODUCTION

This chapter examines the socio-economic issues and local micro-structures that emerge from the generalised political, economic and social issues discussed in chapter 1. Local land users, their decision making and roles in environmental change are enabled and constrained by many socio-economic variables and structures that influence and are influenced by perceptions and access to environmental and human resources and structures. The main topics reviewed in this section are the theoretical and applied issues of livelihoods, household relations, gender relations and the concept of sustainable development. These determine the sustainability of agricultural livelihoods, contributing to land access, local resource management, landuse and general environmental change. The chapter also examines the dominant discourses of development theory and practice, including the contrasted first and second development decades, the rise of environmentalism, the theories and classification of participation, and the applications of these to African examples. These issues are particularly important for the political ecology of soil erosion, as they illustrate the factors for the marginalisation of local actors and their networking, relations, reactions and innovations, which may contribute to soil conservation and/or erosion. Factors for these relations include the links between local agricultural actors and political macro actors, the topic of the next chapter.

2.1. Livelihood Systems.

The concepts of political ecology and structuration theory are applicable to the analysis of livelihood activities, as livelihoods are structures developed by actor decision making, in the face of the wider socio-cultural, economic and environmental structures; and as structures, livelihoods may enable or constrain the activities of the actors. Activities within livelihoods, such as soil management, cultivation and alternatives to farming affect the sustainability of agriculture and possible soil erosion. Agricultural sustainability and soil erosion, as socially constructed environmental and social structures, also constrain and enable actors in their construction of livelihoods, under the wider social and economic structures of the society.

A livelihood may be defined a system of adequate stocks and flows of food, cash, capabilities, assets (income, stores, resources, claims and access) and activities required for daily life (Elliott 1994). A sustainable livelihood is that which can cope with and recover from stress and shock (Chambers and Conway 1992). Livelihoods may exist at several hierarchical levels: individual, household, extended family and social group or community, and may comprise many activities. The scale of livelihood linkages relates to poverty and influences the nature of the variables utilised for its sustenance (Reardon and Vosti 1995). Livelihood analysis is an important building block of the concept of social and environmental coevolution, because individual livelihood activities contribute to environmental change, which affects them (Cline-Cole 1995, Reardon and Vosti 1995). Chambers and Conway (1992:11) distinguish between tangible (stores, resources) and intangible (skills, knowledge) assets for livelihoods. These include technologies, environmental resources, and knowledge systems, all of which are part of the coevolving system of society and environment envisaged by Norgaard (1994). These assets may be utilised to construct a livelihood comprising several activities, which will therefore involve networking within the socio-environmental system and thus have wider ramifications with the political ecology of the context (Clin-Cole 1995, Ngwa 1995). Activities may include cultivation, herding, hunting, gathering, wage labour, trading, artisan work, processing, and provision of services such as transport (Chambers and Conway 1992).

An element of unpredictability always exists in socio-environmental contexts, and is relevant to environmental and social sustainability. Environmental sustainability depends on the tendency of actors to maintain, improve or degrade the local environment (Moock and Rhoades 1992; Woodgate 1992; Holling 1993, Redclift 1993; Tiffen et al. 1994; Ghersi and

Martin 1996; Hakeem and Vemuri 1997; Row 1997; Scoones 1997a). Woodgate (1991) and Reardon and Vosti (1995) give the example of poverty and livelihood constraints causing farmers to extend their activities to fragile lands. Social sustainability concerns the ability of a livelihood to resist or recover from shocks and stresses, which are socio-economic and environmental problems that threaten livelihood stability and existence. These problems may exist at the individual or household level. Examples of stresses are population pressure, unemployment and unfavourable terms of trade. Examples of shocks are fires, wars and sudden drought (Chambers and Conway 1992). Such problems cause change in socio-environmental inter- and intra- relations, influencing the coevolution of structures dependent on multitudes of actor actions.

2.2. THE HOUSEHOLD AND GENDER RELATIONS

The household is an important "building block" of food security, rural socio-economic and environmental relationships, and is therefore an enabling or constraining livelihood structure (FAO 1996; Zelezny et al. 2000). There is a strong, reciprocally supportive relationship between the household and livelihoods. Arguably "ignored in macro-economic analysis" (Floro 1995: 1913) households are defined as sites "within which a wide range of activities - production, consumption, child bearing - are carried out often in vastly different ways" (Crehan 1992: 135). Household rules and relations enable and constrain wider social relations, yet are also enabled and constrained by actor agency and wider social relations. They also "provide a wide array of goods and services for their own use and consumption, including subsistence crops, gathered water and fuel, food preparation and house cleaning, care for the children and elderly, and community management and services provided by voluntary work" (Floro 1995: 1913). The village community is composed of many interlinked households (Gill 1991; Carter 1997). Linkages, which comprise important structures, may take the form of labour exchanges for farm work, employment and financial credit. These linkages help individual farmers and households to cope with environmental and socio-economic stress (Carter 1997) and to develop or adopt appropriate technology (Moock and Rhoades 1992).

The socialisation of children for adult roles is a very important function of the household. Learned culture is a strong force that influences people's perceptions (Girvan 1995; Zanen 1996; Subrahmanian 2005). Geertz (1973:

49) argues that culture is human nature and human nature is culture; therefore, there is no such thing as human nature independent of culture, because culture is the essence of people's existence. Banuri (1990) links cultural structures with the structuration theory of Giddens (1979; 1984) by pointing out that cultural maps are constructed by individuals through their life experiences, while individuals also influence, through social relationships, the cultural maps of other people. The aggregated cultural structures that result also affect the construction of an individual's cultural maps, as they are part of the individual's social reality. Papanek (1990), Vlassoff (1994) and Subramanian (2005) stress the importance of culture in determining perceptions of status and gender relations, and note that more research must be undertaken to investigate the cultural factors that contribute to gender inequality.

A crucial issue concerns the gender specific nature of many household and related livelihood activities and the effect that this has on people's perceptions of gender relations. It may enable some activities through specialisation, but also restricts livelihood choices according to gender and creates structural, gender related differentials in poverty. Differentials in the training of boys and girls may encourage the lower status of girls, and this may be an influence between gender related power relations (Papanek 1990; Vlassoff 1994; Zelezny et al. 2000; Subrahmanian, 2005). Handa (1994) argues that the lower status of women may result in the loss of opportunities for children living in female-headed households. Poverty tends to be gendered (Crehan 1992: 129), in the sense that men may have easier access to employment and resources, due to their early socialisation as leaders, within the larger issue of poverty and environmental degradation (Zinyama 1995).

Studies of household decision making and farming have neglected gender relations (Ellis 1993; Bendini and Palomares 1995; Campbell 1998; Elliott and Campbell 2002; Blackden and Wodon 2006; Food and Agriculture Organisation 2010). There are noticeable exceptions (Shiva 1989; Muylwijk 1996; Muylwijk and Smetsers 1996; Merchant 2005; Nelson and Stathers 2009). Ellis (1993) argues that women are the invisible agricultural producers in peasant society. Taking the household as a generalised unit of analysis may obscure different, gender specific tasks, the effects on production decisions and income distribution within the home (Ellis 1988, 1993). The important issues in gender relations are gender distinctions in labour, land and credit access, income levels, extension contacts and technological developments (Henderson and Hansen 1995; Samanta 1995). The reason for this omission is the focus of agrarian development research on male household heads, an approach that neglects the productive role of women at all stages of the food production process. Nevertheless, Handa (1994) adds that the increasing incidence of female-headed households (FHH) in developing countries has stimulated policy research on the link between the gender of the household head, poverty and the status of children.

Ellis (1993:166) points out that the division of labour between the sexes is not "natural" in the sense of being ordained by biological differences. Jackson (1994) concurs: gender is a culturally specific and socially conditioned identity of men and women - it is not fixed nor is it biologically determined. Peasant households, where women often do onerous physical tasks, with less access to education are living proof (Aikman and Unterhalter 2005; Subramanian (2005). Only childbearing is necessarily a female function. Because the roles of men and women are socially, rather than biologically, determined, Ellis (1993) argues that such arrangements are changeable, and this is the central point of the feminist critique of social science analysis. Nevertheless, men and women have accepted gender differences as a natural corollary of the biological differences between them (Sen and Grown 1987; Peterson and Runyan 2010). The result is that women are constrained in their income earning opportunities and access to resources (Handa 1994; Rocheleau et al. 1996; Aikman and Unterhalter 2005). Therefore, cultural ecofeminism seeks to increase women's involvement in decision-making and also to increase their access and control over resources (Pepper 1996; Dankelman 2010). The arguments for such new perspectives challenge established structuralist positions and reinforce the structuration theory through an emphasis on individual action.

The two main socially determined roles of women in peasant households are household work and production, on and off the farm (Galan 1994; Aguilar 1995; Henderson and Hansen 1995; ILO 1995; Samanta 1995; Peterson and Runyan 2010). Pepper (1996) points out that in the developing world women constitute the backbone of production as well as reproduction. Household work involves the rearing of children, cooking, cleaning, washing, sewing, firewood collection, the fetching of water, household repairs and building. On-farm productive work may include food crop cultivation, animal husbandry and food processing (Bryceson and Howe 1993; Henderson and Hansen 1995; Floro 1995; Samanta 1995; FAO 1996; Grigsby 1996; Jacobs 1996). Off-farm productive work may involve alternative livelihood activities such as trading for extra income. The combination of gender-based work activities and accessibility rights affect work schedules and constrain and enable the success of activities (Ellis 1993; ILO 1995; Rocheleau, Thomas-Slayter and Wangari 1996; Aikman and Unterhalter 2005).

The main issue in these relationships is power, manifested by the dominance that men hold over women in household and related decision-making (Turner and Brownhill 2006). This has pronounced socio-economic implications, even in education (Subrahmanian 2005). A great deal of the work done by women is unpaid labour, supportive of male dominated activities. This fact neglected by socio-economic analyses, which lack gender sensitivity (Galan 1994; Bendini and Palomares 1995). Task allocation structures (allocative resources in Giddens' terms) and gender specificity, which influence both men and women in decision making, are largely created and reinforced by male decision making (Floro 1995; Peterson and Runyan 2010). Ellis (1993) points out that in strongly patriarchal societies, men may exert control over household income, women's working time and freedom of movement, and even their levels of consumption. Technological change in agriculture, therefore, may actually increase gender differences, because men are socially prepared to usurp leadership positions in any livelihood activity (Mayoux 1995; Grigsby 1996; Aikman and Unterhalter 2005; Dankelman 2010).

Although in many societies traditional gender distinctions are breaking down, old patterns may be replaced by new gender differentials as men take over the more attractive roles, leaving women with those that are more arduous or less rewarding (Gill 1991; Aikman and Unterhalter (2005). The introduction of cash cropping and mechanisation are examples: men take over new cash cropping methods and mechanical developments, leaving women to pursue subsistence agriculture (Jacobs 1996; Campbell 1998). Credit facilities may also favour men, hindering an equitable sharing of responsibilities and rewards (Albee 1996). Conversely, with the introduction of new opportunities, individual female actors may seize upon the situation to better their lot (Dankelman 2010). Household welfare is likely more important for women then men (Gill 1991). The gender and other social relations within it are thus important, as well as the wider socio-environmental, resources and relationships within the village, region or state (Aguilar 1995; Floro 1995; Grellier 1995; ILO 1995; Carter 1997; Yang 1997; Banerjee and Mayerfeld Bell 2007) thus influence the strategies that a household may employ to cope with poverty.

In response to shortages, especially those of a seasonal nature, there are several individual coping strategies for households: the diversification of income sources; diversified resource exploitation; increased social networking and contacts; movements; creation of household stores; the reduction of consumption; and the exploitation of assets (Chambers and Conway 1992).

The building blocks of this macro level spatial change relate to individual perceptions of livelihoods and the environment within the household (Campbell 1998). There may also be counter effects, as environmental change may reduce available resources and hence constrain individual decision makers in livelihood activity diversification (Reardon and Vosti 1995; Dankelman 2010). Crucially, these differing approaches to peasant farmer decision making and household relations offer complementary, rather than conflicting, understandings of the reality of actor livelihoods. In many cases, farmer livelihoods may disintegrate in the face of external, constraining pressures; in other cases, the household may be able to cope by adopting strategies that make recourse to the enabling factors that support such activities. Therefore, the important point is the acknowledgement that the subjectivity of actor decision making is highly complex and this complexity encourages different emphases in explanation (Reardon and Vosti 1995).

2.3. Gender and the Environment

Related to the household analysis of gender relations is the substantial body of literature that offers contrasting views regarding gender differentials in environmental attitudes and knowledge (Marchard and Parpart 1995; Buckingham-Hatfield 2000; Merchant 2005; Sweetman and Porter 2005; Banerjee and Mayerfeld Bell 2007; Dankelman 2010). Ideas in ecofeminism, the paradigm that links feminism to ecology, cluster around the thesis that women, culturally and/or naturally, have closer affinity to the natural environment, women and men have different relationships with the environment, and women and the environment experience a similar degree of male dominated exploitation (Mellor 1997; Banerjee and Mayerfeld Bell 2007; Shiva 1989). The ecofeminist perspective is based on the analogous, dualistic symbolism of male dominance over women, society over nature, science over popular knowledge, abstract knowledge over personal and particular knowledge, and macroeconomics over domestic and household relations (Sweetman and Porter 2005; Banerjee and Mayerfeld Bell 2007).

The women's and ecology movements are therefore held to be primarily counter trends to a patriarchal mal-development. The two key arguments that underlie this conception are: (1) that women have a special and close relationship with nature; and (2) that women are particularly sympathetic in their environmental management, forcing the conclusion that women are the natural supporters of socio-environmental development (Jackson 1993).

Some scholars (Agarwal 1991; Rao 1991; Mellor 1997) reject the ecofeminist view that women are the natural managers of ecosystems, or are culturally closer to nature because men have distanced themselves from nature. Merchant (2005) argues that this special relationship of women to nature and politics makes it difficult to envisage that men can also posses an ethic of caring for nature. As pointed out by Jackson (1993), rational short term interest may provide a better explanation for women's environment - friendly management practices than the implicit assumption that women are caring, nurturing and selfless beings, committed to both future generations and to the environment for its own sake.

A further argument claims that assumptions about the complementarity of women's interests with those of the environment are flawed, the result of a failure of both ecofeminism and technocentric environmentalism to comprehend the conflicting interests of women and environments. This would be because of ethnocentrism, essentialism and gender blindness, as well as a lack of social, historical and material analysis (Jackson (1994). Therefore, environmental conservationism and gender identity are not necessarily parallel, but are contextual variables, resting on the livelihood and the class/gender system. For example, within the livelihood system, the source of basic needs and economic resources perceived by women as intrinsic to their culturally determined position naturally becomes a focus for their attention (Grellier 1995; ILO 1995). This may be a socio-economic priority, rather than an environmental interest (Henderson and Hansen 1995; Samanta 1995; Buckingham-Hatfield, 2000). Also, the class/gender systems, which define the status and role of women in societies, may influence their environmental outlook (Mellor 1997). Those in the upper classes, who are not conditioned by upbringing and context to regard the environment as the source of their livelihood, do not necessarily view the environment with any more favour than men.

The dynamic nature of agro-environmental relations may also influence a modification, through time, of women's environmental relations (Dankelman 2010). A case in point is the impact of male migration on the agricultural role of women and the female role in forest utilisation. The assumption in this case must be necessity and survival (Bliss 1996; Grigsby 1996, Marks 1996). The increasing incidence of female migration, which results from the more "modern" outlook of the younger generation, points to a similar desire among women and men, and an increasing rejection, among the women, of their cultural role as sustainers of life, whether or not this benefits the environment (Campbell 1998). Jackson (1993) therefore argues for the replacement of the assumption of intrinsic environmental commitment among women by a more realistic approach, especially among development agents, which will endeavour to build stronger incentives for women land users for land conservation.

These arguments provide the basis for gender analysis (Razavi and Miller 1995). This holds that men and women's environmental knowledge are similar and that women have no inherent closeness to their environments (Jackson 1993). There should be a gendered analysis of property relations as a precondition for understanding the differential incentives of men and women toward conservation. Gendered environmental decision-making needs to be seen through the lens of intra-household relations of both co-operation and conflict. Gender analysis focuses on gender relations, rather than the concept of patriarchy and the critique of western rationality emphasised in ecofeminism. The notion of socially constructed and variable gender identity is central in gender analysis, rather than the existence of women as a social category that is fundamental to feminism. Proponents of gender analysis argue that it is not enough to examine the concept of patriarchy in a generalised, monolithic, ahistorical and reductionist manner; there must be an investigation of the social dynamics that originate, reproduce and change such structures (Jackson 1994). Clearly then, Jackson's approach owes much to structuration theory and complements the model of political ecology. These ideas with household relations and livelihood structures form the basis for sustainable development.

2.4. The Paradigm of Sustainable Development

Sustainable development was first mentioned in the World Conservation Strategy, and from there evolved into a core modern concept, policy objective and principle for policy intervention relevant to all strands of environmental management (I.U.C.N. 1980; Adams 1990). It became a fashionable theme after the World Commission on Environment and Development ("the Brundtland Commission" 1987) expressed it as development that "meets the needs of present generations without compromising the ability of future generations to meet their own needs" (Serageldin 1993; Soussan 1993). The concept has since attracted the attention of scholars from many disciplines and policy orientations. Sustainable development is a concept that evolved from an emphasis on ecological problems (I.U.C.N. 1980), to include economic problems (Brundtland Commission 1987; Soussan 1993). It is also a principle for policy intervention, exemplified by "debt for nature swaps" or the environmental conditionality of aid (Jacobs 1991). Sustainable development is also a political objective, as governments are under pressure to demonstrate their concern for socio-environmental issues to pressure groups, the voting public and other governments. However, there is no consensus on the method and goals of sustainable development (Redclift 1987; Pearce et al. 1990; Serageldin 1993; Ingham 1993; Soussan 1993). Economists emphasize the maximisation of human welfare within the constraints of natural capital stock and technology (Common and Stagl 2005). Ecologists aim to preserve ecosystems from a physical, not monetary, perspective, to ensure the overall stability of the global ecosystem (Serageldin 1993; Apostolopoulou and Pantis 2010). Radical or "dark green" political thought emphasises a limits to growth approach, stressing a human partnership with nature, and the limits the finite "carrying capacity" of the Earth places on economic, technological and population growth (Dobson 1990; Irvine and Ponton 1988; Trainer 2010a).

Some economists perceive sustainable development as the maximisation of the net benefits of economic development, subject to the maintenance of services and the quality of resources over time. Thus, the rate of utilisation of renewable resources must be equal to or less than natural replenishment, and non-renewable resources require effective management (Pearce and Turner 1990; Common and Stagl 2005). This optimistic view of sustainable resource exploitation has not always been the sole perspective of economists. The early approach of economists like Malthus (1798) was pessimism about the sustainability of long-term economic growth. Diminishing returns would ensure that as population grew and land remained fixed, scarcity would result. John Stuart Mill saw economic progress as a race between technical change and diminishing returns in agriculture. Marx held that capitalist economic systems were inherently unsustainable. While there would be some short-term advantages, like labour saving devices, in the long term there would be negative effects, such as pollution and environmental degradation. These would be the result of an excessively exploitative political and socio-economic system (Merchant 2005).

Despite a growing interest in environmental issues during the 1970s, many economists saw economic growth as feasible and desirable. Technical change would counter resource depletion, improve the quality of labour and capital, and allow continued exploitation of resources (Pearce and Turner 1990; Common 1988; Campbell 2005a). This approach is idealistic because it assumes a technical rationality of actors for the optimal and efficient exploitation of resources. Economists argue that a properly functioning price mechanism can ensure environmental sustainability; a view Common (1988) terms the "simplified position of mainstream economics." The over-exploitation and consequent scarcity of a resource is supposed to cause a price rise, and a reduction in consumption and production. The system will shift away from environmentally damaging methods, and virgin resources will become increasingly expensive, encouraging recycling (Common 1988; Common and Stagl 2005).

Opposing views emerged in the 1970s, because of environmental problems, the increase of the oil prices, the publication of the Limits to Growth thesis (Meadows et al 1972), and the failure of the economic growth oriented First Development Decade (Todaro 1989; Campbell 1998). A new "Basic Needs Approach" based on the reduction of poverty, inequality and unemployment developed among policy makers as the answer to development problems (Seers 1982). Daly (1980) argued for the abandonment of the growth objective, to ensure environmental sustainability. Important new arguments supported the least advantaged in society (intergenerational equity) and for fair treatment for future generations (intergenerational equity) (Pearce et al. 1989a; 1990b; Rawls 1972). Munasinghe (1993) points out that economic optimisation and efficient resource allocation are difficult to apply to some environmental and social objectives. These include the preservation of the resilience of ecosystems, the promotion of public participation and the reduction of conflicts. However, in spite of these arguments, mainstream economics still points to sustainable economic growth as a possibility, given optimal and efficient utilisation and maintenance of the natural capital stock (Common 1988).

This view is a sharp contrast to the radical Green position. A key Green argument is the advocacy of "zero growth" (Trainer 2010a, 2010b). Contrary to the modern "technocentric" cultural viewpoint, that emphasises technological solutions to socio-economic and environmental problems O'Riordan 1989; Bowlby and Lowe 1992), Greens hold that a change in human values and morality is necessary to reduce excessive consumption patterns (Irvine and Ponton 1988). Economic growth requires cancellation, rather than modification, because it is the primary cause of environmental degradation (Dobson 1990; Jacobs 1991). The Earth is limited in resources, hence the exponential rate of increase of industrial growth is limited (Irvine and Ponton 1988; Campbell 1998). The ecological health of the Earth determines human welfare. Non-human life is as important as human life and has an inherent right to exist; it is not merely natural capital stock. Humans are capable of a high quality of life independent of economic and material well being (O'Riordan 1981; Ophuls 1977; Bunyard and Morgan-Grenville 1987).

Jacobs (1991) argues that a fundamental weakness of the Green position is its failure to distinguish between different meanings of economic growth.

Greens take economic growth to mean an increase in consumption, while economists take a broader view, measuring income flows within the society, and incorporating activities that may or may not damage the environment. Thus, it depends on which variable is growing. In addition, the "zero growth" system does not take into account distributive issues; environmentally harmful activities may differ according to income level (Jacobs 1991). Nevertheless, Greens point out that the economists' obsession with the Gross National Product and economic growth reveals a system that neglects future generations, perceives the Earth as a resource rather than a blessing, "produces indiscriminately and consumes voraciously" (Dobson 1990). Greens favour political decentralization and other local actions, while emphasising the need to think globally. They call for ecocentric policies, labour intensive modes of production, individual self-sufficiency, cultural diversity, fair distribution of wealth, reduced consumption and lower population levels (Ophuls 1977). There is however disagreement over the means of reducing the human population. Some radical Greens, such as the Earth First Group, have suggested that diseases should be allowed to increase mortality. Others, such as the British Green Party, reject "unpalatable legislation" (Irvine and Ponton 1988), and favour negotiation (Bunyard and Morgan-Grenville 1987; Dobson 1990).

The Green position on technology is unclear. Technology is behind many environmental problems. However, they welcome appropriate technology that is sensitive to environmental and social concerns. Examples of such technology are labour intensive agriculture, and recycling devices (Dobson 1990). Trade can operate against sustainable self-reliance and encourages domination. Protectionism could help (Dobson 1990), but Brundtland (1989) argues that protectionism is an aspect of domination between nations and should be abolished.

In contrast to the sharp and radical arguments offered by the Greens, ecologists focus on the impact of development policies on natural ecosystems (Apostolopoulou and Pantis, 2010). They encourage the integration of ecological considerations into development policies and the formulation of anticipatory and preventive strategies for projects (Rees 1993). A key problem facing ecologists is to demonstrate that they are not just "a lot of bird watchers" (Boulding 1966). They must convince policy makers that environmental issues complement development, rather than hinder it. In the pursuit of this goal, ecologists must work closely with economists and politicians to devise regulatory, management and social measures to ensure environmentally sustainable development. Recent developments include the adoption of environmental action plans, environmental impact assessments on proposed projects and the encouragement of sustainable agricultural practices (Ree 1993; Campbell 1998).

Tisdell (1990) argues that ecologists have an unclear conception of sustainable development. Nevertheless, a key theme that emerges in the writings of ecologists is the maintenance of biological diversity and productivity, and the resilience of ecosystems to external influences (Simmons 1974; World Conservation Strategy I.U.C.N. 1980; Conway 1983a, 1983b, 1985a, 1985b; World Commission on Environment and Development 1987; Riddell 1981; Campbell 1998; Apostolopoulou and Pantis, 2010). Natural methods are preferable to "artificial" methods of environmental management. Thus, natural predators can substitute for pesticides in pest control, and multiple cropping and agroforestry can help crop output and maintain ecosystems. Biological diversity maintains ecosystems. Thus, ecologists attack those human activities that reduce biodiversity (Apostolopoulou and Pantis 2010). The scientific, factual, ecological perspective provides a background for Green principles (Irvine and Ponton 1988; Yearley 1991). Arguably, the ecologists' approach puts the environment first, rather than the people first. They seek to protect the environment from people, even when it is people and not the environment who are vulnerable (Davies 1992). Poor people, especially in the developing nations, depend on the environment for survival (Campbell 2005b). Therefore, social factors are important. However, ecologists may argue that environmental protection benefits everybody, because degradation will affect the poor as well as the rich. They suggest that while social factors are important, effective environment management is the key to sustainable development.

These broad, differing conceptualisations of the concept of sustainable development underlie attempts to develop it as a policy intervention. In the 1980s, aid flows from the developed to the developing nations were increasingly subjected to "Green conditionalities" (Davies 1992). Environmental sustainability was taken as the guiding principle behind some financial and technological assistance. Donors insisted that governments adopt environmentally sustainable policies as a condition for aid. Environment issues claimed aid otherwise intended for other projects For example, renewable natural resources took 14 percent (£37 million) of the O.D.A.'s financial aid in 1980, 27 percent (£60 million) in 1987 and 22 percent (£65 million) in 1988 (Davies 1992). Many environmental priorities were set in the north, and hence did not fully consider socio-cultural and political contexts in the developing nations (Davies 1992). In addition to "green aid", many governments, corporations and organisations were under pressure to adopt environmental policies (Yearley 1991). To appease green pressure groups, the voting public and other activists, governments took a tougher stance on environmentally degrading practices. Soussan (1992) notes, because of political pressure, politicians may ignore such issues "at their peril." Action, however, was rare. Powerful stakeholders, ineffective legislation and insufficient funding posed problems.

Despite its importance and near universal real and superficial support, sustainable development, its methods and aims, did not, and does not enjoy a unified definition or perception. This represented the failure of contemporary thought to define the exact role of the environment within the devised development paradigms. Despite the prevailing disputes over what constitutes optimum development, a positive issue to emerge from these debates and experiments is the recognition of the importance of a broad integrated approach for the investigation of development and environmental problems. This is especially relevant in the case of agricultural development, the understanding of which necessitates the investigation of historical, political, economic, socio-cultural and environmental variables.

2.5. The Political Ecology and History of Agricultural Development

The tenets of sustainable development, viewed variably by various and sometimes-opposed stakeholders have been relevant to agriculture, possibly since the development of settled cultivation. All the three definitions of sustainable development discussed above are relevant to agriculture, as agricultural development is an economic, ecological and green issue. Currently, the historical contextualisation of resource access and the consequences for agrarian economics, green socio-environmental relations and environmental conservation are important topics (Gadgil and Guha 1992a, 1992b; Peluso 1992; Little 1992; Bassett and Crummey 1993; Faber 1993; Agarwal 1994; Fairhead and Leach 1994, 1995, 1996; Norgaard 1994; Peters 1994; Painter and Durham 1995; Moore 1996; Campbell 2004, 2005a, b).

Within this broad discourse, political and agro-ecological issues are interlinked. Burke (1984: 191) describes politics as the "child" of agriculture because the increasingly complex environmental modifications that began to form the basis of human existence brought complex questions and issues into the human consciousness. These included the need for territorial definition and the development of mechanisms for the defence of such resources in the face of competition from other similarly developed polities. Land access was important, and evolved into a definer of power among classes and actors in the society. Differentials in soil fertility and water, and crop suitability created conflicts and raised questions of allocative justice and power, with marginal lands usually assigned to marginal actors. The necessities of management, storage and distribution of surplus production raised questions of effective resource management and distributive justice. Intra-societal, as well as inter-societal conflicts became vital aspects of the new reality. Political distributive systems became the bases for agricultural decision-making and access to environmental systems.

Different state agricultural policies have been debated as possible solutions and causes of agricultural and environmental problems. The impacts of such policies on population-environment relations are as diverse and dynamic as the motivations for undertaking them (Elliot 1995). The increasing realisation of the importance of political ideologies, economic development models and social perceptions in the design of resource exploitation systems, has led to a succession of development paradigms, each supported by an extensive literature of varying relevance and impact on agro-ecosystems (Peet and Watts 1996). During the 1960s and the early 1970s (the period of the first United Nations Development Decade) the dominant idea shared by donor agencies and policy makers was the definition of development as a scientific exercise, with rational, comprehensive planning and measurable factors of success (economic, measureable by per capita income and consumption, the gross domestic product and the area under cultivation) (Todaro 1989; Hatzius 1996b; Peet and Watts 1996). This represented what was termed the Dominant Development Paradigm (Brock 1994). This ignored or marginalised social concerns (such as poverty, inequality, unemployment and gender differentials) and environmental issues (such as deforestation, pollution, the loss of biodiversity and soil erosion).

The failure of the First Development Decade to serve as the genesis of "Third World" development and the serious environmental and social dislocations that some of its policies induced, led to a revision of development related thinking in the 1970s. The product was the Second Development Decade, with the emphasis centred on the reduction or prevention of poverty, inequality and unemployment and a greater recognition of the importance of environmental issues (Hope 1984; Todaro 1989; Stern 1996). This shift in development thinking was particularly relevant to the core issue of agricultural development and consequent environmental change, because the exploitation of environmental resources to support the economic growth (or dominant development) paradigm was a major contributory factor behind the environmental disasters (especially soil erosion, drought, desertification and famine) which occurred in Africa during the 1980s (Elliot 1994; Campbell 1998).

However, despite the more humanised focus of the second Development Decade, the problems concerning rural development were not solved (Stern 1996). Schuurman (1993) notes that the 1980s were a lost decade to both developing countries and development theory. There was also a growing awareness of an impasse between development theory and reality (Booth 1993; Schuurman 1993). Development theorists and practitioners began to realise that development theory had become irrelevant to the needs of the objects of its study (the poor in developing countries). The economic growth focus of development theory contributed to this impasse: it created a theoretical focus that was largely insensitive to the natural environment. The global collapse of Marxism was also a causative factor, as socialism was questioned as a development model, and many Marxist approaches were also perceived as generalised, economic and structuralist, neglecting, individual agency, culture and micro-studies (Booth 1993; Peet and Watts 1996).

The emergence of "environmentalism" and a Green movement as contemporary intellectual and political phenomena was a major contributory factor behind the revised perceptions of agro-ecosystems and environmental resources that occurred during the 1970s (Martinez-Alier 1992; Sachs 1992; Guha 1994; Yearley 1994). Terms such as sustainable agriculture, agroecology, and organic farming became fashionable "battlecries" and value representations for a new politically and ecologically based agro-environmental awareness, within the grammar of sustainable development (Escobar 1996). Sustainability and environmentalism acquired increasingly broad perspectives (Buttel 1992; O'Connor 1994), giving a fuller understanding of the multiplicity of ways of comprehending the extraordinarily complex nexus of development-environment relations (Peet and Watts 1996: 38). The Stockholm Conference of 1972 was an expression of the international environmental concerns developed over the previous decade. Pronounced institutional modifications in many countries reflected the impacts of these issues. By 1974, more than one hundred countries had developed environmental institutions. Environmentally biased non-governmental organisations were also set up and many new environmental laws adopted (Elliot 1994; Convery 1995).

Most of these efforts however were representative of the conventional ecological narrative and a fundamental problem was the failure of environmental protagonists to enlighten policy makers on the correlation and coevolutionary relationship between agricultural development and socio-environmental issues (Yearley 1994). The idea that environmental conservation was in conflict with the needs of agricultural development clouded the perceptions of policy makers throughout the international economic system. Thus, while environmental issues received "lip service", effective action was rarer (Finger and Verlaan (1995); Eyzaguirre 1996; Monu 1997). Nevertheless, scepticism about modernity and its rationality developed, and a fundamental position of these competing discourses was a focus on the problems of economic and technical development (Peet and Watts 1996; Boughton et al. 1995; International Research Institute 1995).

For example, Ellis (1993) argued that large scale, highly technical and capital-intensive farming was unsustainable compared with small-scale farming. His thesis rested on the proposition that large-scale agricultural ventures created various opposing forces degrading to peasant livelihoods and farm production. There might be small farmer debt, distress sales of land for income and migration to wage labour on larger enterprises. These factors might sustain, modify or degrade the actor's livelihood. Agrarian sustainability would rest on the reliability of land accessibility, market connections, social networks and market linkages. Other trends such as increased specialisation, stronger dependence on market linkages and greater competition could influence farmers to set up family farm enterprises (Ellis 1993).

Practically, there is evidence that large scale "industrialised" agricultural operations have a poor performance record (Johnson and Ruttan 1994: 691). Reasons advanced for this trend included: the tendency for enterprises to focus on technical and economic viability rather than socio-political variables; ecologically inappropriate mechanisation methods; failure to consider socio-environmental contexts; high costs of inputs; and ineffective marketing and distribution systems (Johnson and Ruttan 1994; Rukuni 1995). Therefore, while accepting that there are exceptions to these ideas, the sustainability of large scale, industrial farms in locally variable contexts is suspect (Johnson and Ruttan 1994; Rukuni 1995).

Irrigation projects are good examples of large-scale economic endeavours that may fail due to insufficient consideration of socio-environmental contexts (Rukuni 1995; Ciancaglini 1996). Wickham (1986) argues that irrigation is an important modification in areas of insufficient rainfall. Most such areas were savannas, where there was a prevailing belief that the solution of the low rainfall 'problem' was irrigation (Rukuni 1995). Many governments were pessimistic about crop production and savanna problems (Young and Solbrig 1993). Problems that hindered these enterprises included outdated administrative methods, poor design (especially of drainage networks), generalised approaches to operations, poor day-to-day management and inadequate participation by local people (Rukuni 1995; Ciancaglini 1996). Further, there was a neglect of how gender relations may enable or constrain the adoption of irrigation technology, due to differences in male and female socio-economic interests, perceptions of user costs, and long and short-term benefits (Zwarteveen 1994). Agricultural extension and information utilisation was frequently inefficient due to insufficient numbers of officials and insufficient consideration of different approaches (Adimorah 1995; Niekerk 1995; Ban and Hawkins 1996; Christoplos 1996; Monu 1997; Platteau 1997; Wallace 1997). The sociological contexts of irrigation and other projects were therefore important foci for the success of irrigation (Ciancaglini 1996; Monu 1997).

CONCLUSION

The main theme of this chapter was the rethink of nature-society relations, acknowledging the contribution of local microstructures, such as livelihoods, households, and gender relations to rural development. These ideas operate within the paradigm of sustainable development, which in itself is variably defined. All of the discussed points however emerge from the powerful stakeholder groups: natural and social scientists, green protagonists, policy makers and historians. What about the poor local actor? Their inclusion creates questions of participation. What type of participation would be relevant and effective, what would be the format of such an interaction and what consequences? These questions are tackled in the next chapter.

Chapter 3

STAKEHOLDERS, PARTICIPATION AND DEVELOPMENT

3.0. INTRODUCTION

This chapter examines the theory and application of participation. Participation follows logically from discussions of political ecology, social theory, socio-economic structures and sustainable development, because the principles abstracted from the theories must be implemented with the participation of local actors. Important issues concern the structures that emerge from, contribute to and influence the relationship between macro-stakeholders, be they economic planners, business managers, policy makers, large landowners or even traditional elites, and the local micro-actors who construct local livelihoods, households and gender relations. The chapter discusses participation by degree of local actor empowerment, and lays out its merits and applications. This topic is very relevant to soil erosion, as the local people either participate in schemes to increase production and ameliorate environmental degradation, or may be excluded from participation and marginalised to fragile lands, exacerbating land degradation. In such cases, participation is both part of the prevention and part of the cure. These issues vary according to the character of the physical landscape, which is examined in the next chapter.

3.1. JUSTIFICATIONS FOR PARTICIPATION

The theme of participation is important in agricultural projects (Grellier 1995; Hochet and Aliba 1995; Heck 1996; Peet and Watts 1996; Teme et al. 1996; Witcombe and Joshi 1996; Witcombe et al. 1996; Baidu Forson 1997; Bauer and Hoffmann 1997; Pretty and Shah 1997; Riley and Alexander 1997; Campbell 2005a). The ideal types of participation for local rural development are socio-economic empowerment (Zimmerer (1994) and self-mobilisation (Pretty 1995). These refer to the decentralisation of real power to local actors. Pretty argues that people should be allowed to take independent action to solve local problems and exercise control over decision making, even though they may be supported by requested aid agencies, and may still require the support of an enabling framework provided by policy or other institutions.

In addition, the capacity of local people to resist inappropriate or exploitative strategies from above is an important determinant of their empowerment. Mindful of the marginalised position of most rural decision makers, Scott (1985: 1) calls "strategies of resistance" the weapons of the weak. He lists several strategies used by local people to resist external intervention. These include "foot dragging, dissimulation, false compliance, feigned ignorance, desertion, pilfering, smuggling, poaching, arson, slander, sabotage, anonymous threats, surreptitious assault and murder" (Scott 1989: 5). These methods are covert, to avoid open confrontation, due to the possibility of harsh reprisals from the authorities. However, it is questionable whether these strategies transform the conditions that produce poverty and powerlessness, or just make them more bearable (Johnson 1992: 289).

Several justifications for local participation exist. Firstly, they are unavoidable, as the majority of people in local agrarian contexts must be represented. Hence, there must be a degree of populism (Sevilla-Guzman and Woodgate 1997). Richards (1985: 15) terms the "small farmer" or participatory, perhaps socialist approach to agricultural development policy "agrarian populism." Scoones and Thompson (1994) describe this approach as a populist philosophy and renewed interest in it as a paradigm shift. The tools of this approach are Indigenous Technical Knowledge (ITK), which was later termed cultural knowledge (Mock and Rhoades 1992) and then Rural People's Knowledge (RPK) (Scoones and Thompson, 1994). Populism is relevant in Africa because of the food crises and the failure of many technological injections. Certain researchers attempt to move beyond this approach by emphasising more dynamism in actor differences, interests and qualitative research methodologies (Scoones and Thompson 1994).

Secondly, local actors are good enough. A strong justification for the participatory approach is the argument that local people are capable thinkers who in many cases are able to think through problems and operationalise solutions (Birgegard 1993; Long and Van der Ploeg 1994; Fiege 1995; Tognetti 1995; Darkoh 1996; Heck 1996; Baidu Forson 1997; Bauer and Hoffman 1997; Fujisaka 1997; Monu 1997). Local people may therefore be competent socio-environmental managers, and in many cases they may be "reforming, adapting and managing modernisation" (Bebbington 1996). Zimmerer (1994) points out that social empowerment of local people through democratic participation, decentralised decision making, local economic growth, skills and knowledge development is emphasised in the new ecology paradigm. This point is also explored by several authors (Herlihy 1989; Bebbington 1991; Carney 1991; Nietschmann 1991; Friedman 1992; Peet and Watts 1996; Rowe 1997).

Thomas-Slayter (1994: 1480) argues that over time analysts have observed that project performance is more effective and sustainable development more likely to occur when local residents are active participants in identifying needs and implementing projects for their communities. Referring to the role of small farmers, Richards (1985) notes that they are self reliant and capable of organising the activities necessary for their survival, despite the problems that are largely caused by failures of technical developments. Participatory methodologies in agricultural development are thus the best state investment policies, if they emphasise genuine participation. Any other approach may either utilise the wrong inputs or help the wrong people. Other important participants include government agencies and possibly nongovernmental organisations, but without local people institutional efforts are less likely to succeed (Karanth 1995; Alsop et al. 1996; Pretty and Shah 1997). Pretty (1994) also points out that sustainable agriculture needs: agricultural professionals willing and able to learn from farmers and other stakeholders; supporting external institutions; local groups and institutions capable of managing resources effectively: and above all, agricultural policies that support these features. This requires more research on participatory methods (Aguilar 1995; Bauer and Hoffman 1997; Fujisaka 1997; Monu 1997; Riley and Alexander 1997).

Thirdly, external stakeholders frequently misunderstand the local context. Many policy actors and actions do not change or understand social relations, especially class differences related to finance, marketing and land accessibility (Crow 1992). One reason is the lack of government commitment to agricultural development and to the understanding of the impacts of land reform policies (Elliot 1995). In some cases, state power may support local elites, which causes further differentiation in social class structures. The reluctance of central governments to devolve real power to local people hampers the solution of problems in local contexts. The failure of the state justifies the organisation and mobilisation of rural people, perceiving them as having initiative, adaptability and intelligence (Tiffen et al. 1994). It also assumes that local people's knowledge, values and culture, based on intimate experience of the local context, are forces to be utilised productively. Political empowerment may thus be the means by which local people may put forward ideas for their own benefit (Johnson 1992; Allen and Thomas 1992; Mayoux 1995).

Fourth, smaller farmers or peasants are usually unable to benefit from the large-scale agricultural policy initiatives, despite their importance in food production in Africa and in many developing countries worldwide (Binns 1994; Elliot 1994, 1995; Aguilar 1995; Zinyama 1995; FAO 1996; Hambly 1996; Pretty and Shah 1997). Access to land, credit, labour and technical resources are serious issues in the decision making of poor farmers (Maass 1995). With few sources of credit, they rely on local, informal moneylenders, which creates exploitative relationships, because there are few or no rules that constrain the power of the giver over the recipient. With limited access to finance or participatory approaches to agricultural development, peasants are rarely able to benefit from technical improvements (Ellis 1993; FAO 1996). Furthermore, cultural structures, which may be vital to farmer survival strategies, environmental management and livelihood planning are often neglected or even destroyed in large-scale developments intent on more "civilised developments" (Young and Solbrig 1993: 338).

3.2. Types of Participation

Crucially, not all forms of participation represent the socio-economic empowerment envisaged by Zimmerer (1994: 118), while some technological developments emphasising participation may only succeed through coercion of local farmers (Pretty and Shah 1997). Furthermore, research on participatory methods may be counter-productive if the farmers are not motivated to give of their ideas fully (Fujisaka 1997). Pretty (1995) identifies seven forms of participation as being in common usage. The first four, described as manipulative, passive, consultative and material incentive based, are in reality non-participation and have no positive, lasting effect on people's lives. In manipulative and passive participation, participation is simply a pretence as local people or their representatives have no real power. Consultative participation involves an agenda set, defined and controlled by professionals, without the devolution of real consultative power to local respondents. Incentive-based participation occurs with rewards to participants. However, when the rewards or external pressure cease, so does the co-operation and vice-versa (Pretty 1995; Pretty and Shah 1997).

The last three, arguably better forms of participation advocated by Pretty (1995) are: functional participation, which involves local people in meeting objectives, and may or may not involve them fully in decision making; interactive participation, where local people participate in joint analysis and development; and finally self-mobilisation which refers to participation where people take independent action to solve local problems and exercise control over decision making (even though they may be supported by requested aid agencies). These forms of participation, which depend on political will to enable their successful conclusion, illustrate the complexity of the participation problem (Pretty and Shah 1997).

3.3. Problems Implementing Participation

The empowerment of local people faces many problems, one being the reluctance of powerful elites to devolve power. Chambers (1994b) argues for the status of participation as a philosophy and model in development, but the gap remains wide between fashionable rhetoric and field reality. The bridging of the gaps between local people and development professionals is a problematic endeavour (Scoones and Thompson 1994). Other problems that occur in the empowerment of rural people include the fact that many rural people have differences that may supersede a perception of stakes or a similarity of life worlds. Therefore, communication, the understanding of varying viewpoints, cultural structures, long and short-term interests, the actual possibilities available for individuals to change their livelihoods and possibilities for outside assistance must all be emphasised (Johnson 1992; Scoones and Thompson 1994).

There may also be a conscious bias by the state towards large commercial farms (Ellis (1993). This position is echoed by Pretty (1995: 1252) who points out that farmer initiated mobilisation may or may not challenge existing distributions of wealth and power. Also, Zinyama (1995: 215) describes the wide margin between the policy statements and actual implementation.

This is particularly relevant, as the efficiency of small farmer enterprises is a contested issue in the literature (Barrett 1997). Small farm development under small-scale farmers can be more difficult than that of large-scale farms, with problems concerning coping strategies at the local level and implications for livelihood sustainability and food security (FAO 1996; Monu 1997). People's values and perceptions of enabling and constraining features in their local context affect their coping strategies, both in the short and long term. Johnson (1992) points out those short-term survival strategies may exacerbate long-term problems in the particular contexts. For example, the deforestation of new areas for alternative livelihood resources, or increases in the birth rate to produce more agricultural labourers, may solve some problems in the short term, but create others in the long or term. In addition, in some semi-arid areas even credit facilities may create debts, due to drought (Tiffen et al. 1994). Therefore, local empowerment is not necessarily a panacea for development, but must be critically appraised for relevance.

The obvious solution to gaps in the policy implementation process is more research on such issues to develop new understandings as the basis of policy changes (Batterbury et al. 1997; Bauer and Hoffmann 1997; Pretty and Shah 1997). However, the socio-cultural processes of change that influence relations with new developments and the consequences in terms of farmer decision making are poorly understood and rarely documented (Young and Solbrig 1993; Campbell 2005a). The biophysical factors for local people's decision-making are also often neglected (Batterbury et al. 1997). This makes the prediction of policy outcomes difficult, and promotes environmentally degrading, economically inefficient and inequitable outcomes. This reinforces the position that without local community support, no land use or tenure regime is modifiable.

3.4. Participation in the African Savanna

All the points enumerated above are relevant to the African rural context. Young and Solbrig (1993) note that many African savanna and semi-forest cultures have deep and powerful roots and traditions that defy easy change and stress the importance of socio-cultural and socio-environmental variables. To ensure the operational effectiveness of development schemes, impartial methods and institutions are advocated for the adjudication of land use disputes and land access, while encouraging flexibility of local land management (rather than the imposition of new rigid laws), creating laws that are enabling rather than prescriptive (Scoones et al. 1993). These suggestions open questions about the utilisation of traditional institutions and/or the establishment of new structures and conflicts of interest between local and national bodies that may compromise the fairness of policy implementation (Lawry 1990; Toulmin 1990). This is a particularly serious problem, as many savanna-farming contexts exist under complex socio-economic arrangements that have evolved over centuries (Young and Solbrig 1993).

Scoones et al. (1993) describe the development of village level committees in pastoral Sahelian communities as organisations that are given power to make investment and planning decisions, the idea being that local people are better able to determine their needs and know their institutional capabilities. Norman (1997) also gives the example of a local community that successfully established an irrigation project on the shores of Lake Madarounfa in South Central Niger in the Sahel region with sound knowledge of the local socio-environmental and cultural conditions.

Hambly's (1996) example is of grassroots indicators for the measurement of local participation and socio-environmental dynamics, which appraises local capacities for soil fertility and weather monitoring in Kenya and Zimbabwe, key issues being communication and commitment in knowledge bases of local actors and structures. Batterbury (1996) notes in a study of farmers in the Central Plateau region of Burkina Faso that farmers were described as effective planners who networked within the socio-environmental system for livelihood enhancement. Tiffen et al. 1994) noted this in their study of the Machakos reserve in Kenya: the people were seen not as victims of world forces or as passive recipients of government actions, but as people engaged in a generally successful struggle with their physical and socio-political environment, who made choices about the use of their talents and their assets. The people's motivation emanated from their needs, planning acumen and perception of how best to improve their welfare (Tiffen et al. 1994; Mortimore and Tiffen 1995).

Richards and Diemer (1996) describe projects that emphasise the importance of socio-environmental relations as socio-technical hybrids and hybrid socio-technological ensembles. Citing examples from Burkina Faso, Tanzania, Senegal, Sierra Leone and Nigeria, this study investigates the application of methodologies that emphasise strong relationships between the social and technical aspects of project implementation. These illustrate the importance of actors, structures and the interactions between these as dualities and the perception of the environment as a structure. The recommendations were for the intensive participation of local and external expert actors in project implementation, and comprehensive assessment of the environment as a precondition for the success of such ventures.

CONCLUSION

The main points of this chapter are that not only is participation a clear necessity, and there are forms of participation that approach the ideal, but that the achievement of true participation is extremely elusive, perhaps also illusory, unless there is a serious redistribution of power to local actors. In these scenarios, the environmental variables also assume particular importance, because they further complicate an already complicated socio-cultural, political and economic situation. In an African savanna environment, soil type (according to erodibility and fertility), climate (rainfall quantity and type, and temperature), topography and vegetation cover are also very important natural factors, which affect the sustainability of agriculture and livelihoods. The natural and cultural factors of soil erosion are examined in the next chapter.

Chapter 4

Soil Erosion: Natural Factors and Social Externalities

4.0. Introduction

Soil erosion has been termed "one of the most serious environmental and public health issues facing human society" (Pimnetel 2006: 119). This chapter examines the natural causes and effects of soil erosion in West Africa, highlighting the interlinked nature of these sets of variables. The relevance of theoretical approaches, such as the Universal Soil Loss Equation (U.S.L.E.) and the Wind Erosion Equation (W.E.Q.), to the West African context is also examined. In West Africa, agricultural intensification and forestry are the most serious causes of erosion problems. The underlying political, economic and socio-cultural factors are also considered.

4.1. What is Soil Erosion?

Soil erosion is defined as "the movement of soil components, especially surface litter and topsoil, from one place to another" (Miller and Hackett 2011: 296). Two main forces contribute to soil erosion: moving water and wind, and also the strength or weakness of the bonding of the soil particles. Human factors, such as such as inappropriate agricultural practices and inefficient land management may also increase the incidence of soil erosion. Effects include the deterioration of soil structure and consequent soil compaction, loss of water and nutrient retention capacity, reduction in water infiltration rates, increased runoff and erosion, reduction of soil organic matter and watershed loss. Social effects stem largely from the reduced capacity of the soil to support crops. Crops may be washed away, buried or damaged, the loss of nutrients lowers soil fertility, and the rooting depth may be reduced. Externalities include the loss of rural livelihoods and rural to urban migration. Other social effects may arise from the siltation of waterways, reservoirs and dams and the consequent negative effect on hydroelectric facilities, the destruction of roads and the burial of low-lying farmlands.

4.2. The Soil Loss Equations

The main components of water erosion, as pointed out in the Universal Soil Loss Equation (U.S.L.E.) are rainfall erosivity, soil erodibility, slope length, slope steepness, ground cover and conservation practice (Miller and Donahue 1990; Millington 1992). For simplicity, these have separate headings. However, they are interrelated.

Rainfall Erosivity

This is dependent on the kinetic energy released when raindrops strike the soil aggregates. The force is determined by raindrop size and intensity, and the nature of the ground (Millington 1993). West African rainfall is convectional and intensive (Ayoade and Akintola 1982; Barry and Chorley 1982; Campbell 1998). The rainfall may exceed the infiltration capacity of the soil, causing flooding, landslides, erosion, and high sedimentation rates in water bodies. In many tropical climates, about half the total annual rainfall occurs during only about 10 to 20 percent of the rain days (Channa 1968). Water erosion affects nearly 12 percent of the land area of Africa north of the equator (Food and Agricultural Organisation 1979; Campbell 1998). In Samaru in the northern Guinea savanna zone of Nigeria, the annual rainfall of 1100 millimetres may fall in individual storms of 50 millimetres (Kowal and Kassam 1977). Peak intensities of 120 to 160 millimetres per hour also occur. The high intensity of the precipitation is largely due to thunderstorms. Thunderstorms commonly form in areas of local heating (often with a diameter of less than 10 kilometres), which forces warm, moist air to rise, causing condensation and intensive rainfall, for a few hours (Pettesen 1969). Raindrop collision often produces very large raindrops (Nieuwolt 1977). Convectional rainfall causes serious gulley erosion in the Sahel during the much-awaited rains (Lal 1990). In Burkina Faso, 88 percent of the annual erosion in 1956 and 1957 occurred in 14 hours and 6 hours, respectively, and in some case solid particles were transported at least 1.2 metres by splash (Fourier 1967).

Soil Erodibility

This is dependent on the detachability of soil particles, the break-up of soil aggregates and the infiltration capacity of the soil surface. It is influenced by the cohesiveness, texture and structure of the soil (El-Swaify et al. 1982). Fine particles may block the soil pores, creating a compact surface, reducing infiltration, and increasing runoff, resulting in a sealing effect (Wustamidin et al. 1983; Norman et al. 1984). This may reduce the water infiltration rate by as much as 93 percent (Lal 1976). Soil erosion may reduce the water holding capacity of the soil, by removing finer soil particles and organic matter (Middleton 1985). The material removed is often 1.3 to 5 times richer in organic content than the soil left behind (Foth and Schafer 1980). This affects the soil structure, water retention, cation exchange capacity and nutrient level (Ahn 1993).

The main soil orders of West Africa (aridisols, alfisols, entisols, oxisols and ultisols) are highly erodible (Figure 3). In the dry semi desert and savanna, the poorly formed profiles and low organic matter content of the aridisols makes them very erodible. The occasional rains, combined with long periods of dry, windy weather, further increase the erodibility of these soils (Colombari et al. 1984; Middleton 1985). The widespread nomadic herding and overgrazing reduces the already sparse vegetation (Milas 1984), and compacts the soil, reducing the infiltration capacity and increasing runoff (Foth and Schafer 1980; Ahn 1993). Annual soil losses in this zone were estimated at 10 to 200 tons per hectare (Hoogmoed 1986: Lal 1993). A major factor behind the highly erodible nature of West African soils is the presence of kaolinitic clay. This is a 1:1 lattice clay with one silica layer bonded to one aluminium layer. This bond has a low expansive capacity, limiting the ability of the clay (and hence the soil), to expand and absorb water. The very large particles give a reduced surface area, and a relatively low soil plasticity and cohesiveness (Ahn 1993).

In the alfisols and entisols (the dominant soils of the savanna zone) the sparse vegetation cover, cattle grazing and tillage increase soil erodibility (Roose 1986; Gervy 1987). In severe cases of erosion, the topsoil may be destroyed, leaving concretions of iron and bare rock (Foth and Schafer 1980). The oxisols and ultisols of the forest regions are highly erodible due to their highly leached, deep profiles and low organic matter content (Foth and Schafer 1980; Ahn 1993). However, because of the dense cover of vegetation, which tends to protect the soil against erosion, the potential soil erodibility is not always realised (Roose 1983, 1984; Collinet and Valentin 1984; Ledant 1985). Nevertheless, actual annual erosion rates are estimated at over 100 tons per hectare, especially on steep, cultivated slopes (Lal 1993).

Slope Length and Steepness

Erosion increases dramatically on steep, unprotected cropland (Pimental et al. 1993). For example, in Nigeria, cassava cultivation on land with a one percent slope resulted in a soil erosion rate of three tons per hectare per year. However, on slopes of five and fifteen percent, the soil loses increased to 87 tons per hectare per year and 221 tons per hectare per year respectively. At Ibadan, Nigeria, on bare fallow plots with a 15 percent slope, the annual soil loss was as much as 250 tons per hectare. Mulching minimised soil loss (Lal 1976; Roose 1989).

The low structural stability of some soils on slopes, such those of the alfisols of the upland zones of Nigeria, may weaken the resistance of the slope surface to erosion (Kang et al. 1991). Casenare and Valentin (1992) analysed the occurrence of infiltration and runoff in arid and semi-arid areas of Niger, Burkina Faso, Togo, Cote de Ivoire and Cameroon, and noted that the erosive force on slopes is largely due to surface crusting and vegetation cover loss. The reduced infiltration rate destroys the soil surface (as in Mali, where crop production could be increased by up to 90 percent if the soil infiltration rate on soils were increased from 40 percent to 80 percent (Day and Aillery 1988; Valentin and Casenare 1990). Jones (1975) estimated the total length of active gullying in the Jos plateau as about 4,500 miles, totalling a soil loss of about 100,000 tons.

Vegetation Cover

Vegetation protects the soil from erosion by intercepting rainfall and holding the soil with the roots. The best cover for minimum erosion is dense forest or tall thick grass with dead residue ground cover 4 to 5 centimetres thick (Miller and Donahue 1990). Forest and savanna vegetation is a major factor preventing erosion in Niger, Burkina Faso, Togo, Cote de Ivoire and Cameroon (Collinet and Valentin 1984; Casenare and Valentin 1992). Olayide and Falusi (1977) note that in many areas of cleared forest in Nigeria, between 25 percent and 75 percent of the topsoil eroded, with a loss of fertility. Some farmlands were completely ruined and gully erosion was common in the Sahel regions as the sparse vegetation cover gave little protection (Pieri 1992).

The Conservation Practice

Shelterbelts may be grown to prevent both water and wind erosion as in the Majjia valley in Niger, and at Koro, Macina and Timbuktu in Mali (Angstreich (1991). In some cases, to restore the natural vegetation, sophisticated methods must be employed. Cotthem et al. (1991) describe the use of Terracottem (a soil conditioner mixture consisting of water absorbent polymers, growth stimulators, and mineral and organic fertilisers) combined with erosion control measures for restoring the vegetation of the island of Sal (Cape Verde). Other methods include increasing the planting density for greater ground cover, strip cropping, mulching, reducing cultivation or changing the tillage pattern and terracing with walls or banks.

Wind erosion depends on the velocity and turbulence of the wind and the characteristics of the material over which the wind is blowing. The Wind Erosion Equation (W.E.Q.) is stated as $E = f(I, C, K, L, V)$ where E is the potential erosion loss, I represents soil erodibility, C is a local wind erosion climate factor, K is the soil roughness, L is the length of open wind blow and V is the quantity of vegetation cover. The W.E.Q. predicts that wind erosion is most likely in a context with high winds, a dry climate, poorly aggregated, silty, or humus poor soils, a smooth soil surface, large open fields and no vegetative cover. Wind erosion may be prevented or reduced by the planting of shelterbelts (which must be penetratable to avoid turbulence), strip and inter-row cropping (the planting of crops and erosion resistant plants in alternating rows), and mulching (the addition of manures and other organic matter to the soil).

The soil erodibility (I) of West African soils depends largely on the low structural development and cohesiveness of kaolinitic clay dominated soils, such as aridisols, alfisols, oxisols and ultisols (Foth and Schafer 1980; Ahn 1993). The surface roughness (K) of the soil is relevant to soil erosion because smooth surfaces are more susceptible to wind erosion than irregular surfaces. In open areas of the Sahel and Sudan savannas, the surface roughness is variable, due to the break-up of peds by the occasional torrential thunderstorms, the prevalence of lateritic outcrops and bare rock surfaces, and the deposition of dust and sand on the lee of inclines following dust storms. The climate (C) and vegetation (V) factors are important. High wind speeds during line squalls and drought increase the rate of soil erosion because the low moisture level lowers the cohesiveness of the soil. The Sudano-Sahelian region has experienced major droughts since 1968 (Derrick 1984; Milas 1984). Middleton (1985) attributes the increased occurrence of severe dust occurrences in northern Nigeria, and the six-fold increase in dust storms in Mauritania since 1968 to the cycle of wind erosion, drought and erosion. The length of open wind blow (L) is relevant, because the movement of saltating grains tends to increase with distance. In the Sahel and Sudan savanna, the sparse vegetation enable wind erosion. In the forest, the vegetation is usually sufficient to prevent wind erosion. Therefore, the vegetation cover index (V) is important. Locust swarms, mainly prevalent in the Sahel region, affect the distribution of vegetation cover. Swarms of locusts may be as much as 40 kilometres by 6 kilometres in area, and can consume very large amounts of vegetation (Barrows 1991). Since 1986, locust swarms have increased in Mali, Senegal, and Niger, largely due to the decline of control measures (Barrows 1991).

4.3. The Global Problem of Soil Erosion

Over the millennia, soil erosion has probably destroyed as much arable land as is now cultivated (Kodva 1977). Intensive agriculture is a major causative and contributory factor. Eroded sediments carried by rivers into the oceans increased from about ten billion tons annually before the introduction of intensive agriculture, to about twenty five to fifty billion tons annually, afterwards (Judson 1981). Since the beginning of settled agriculture, about 430 million hectares of productive land has been destroyed (Kodva 1983). Soil erosion has caused many ancient civilisations to vanish (Olson 1981). Deforestation, overgrazing and inappropriate agricultural practices destroyed the ancient Phoenician civilisation, and reduced the Roman grain supply (Eckholm 1976). The ancient kingdoms of Mesopotamia, Syria, Lebanon, Lydia, and northwestern Africa were also ruined by soil erosion and desertification (Lowdermilk 1953; Beasley 1972; Olson 1981).The Harapan-Kalibangan civilisation which existed in the fertile river valleys of northwest India about four thousand years ago was destroyed by soil erosion (Singh 1982). The Mayan civilisation collapsed in 900 A.D., largely due to soil erosion (Olson 1981). The neglect of the elaborate terracing and soil management schemes affected the Inca civilisation of (Denevan 1985; Williams 1987).

Soil erosion has extensive effects on the global social and natural environment (Lal 1988; Millington et al.1989; Goudie 1990; Millington 1992; Wellens and Millington 1992). Buringh (1981) put the annual global loss of agricultural land at five million hectares, due to soil erosion and desertification. A bleaker position is held by Dudal (1981), who wrote that there might be a permanent loss of about six million hectares of productive land annually. Higgins et al. (1982) predicted a twenty-nine percent decrease in crop yields in rainfed areas worldwide over the next twenty-five years. The Food and Agriculture Organisation of the United Nations (F.A.O. 1980) pointed out that soil erosion, along with salinisation, alkalinisation, pollution and nutrient depletion, could eventually lead to the loss of 544 million hectares of cropland.

A major issue concerns the rising world population, and the difficulty of the soil to support increasingly intensive agricultural practices designed to support greater numbers of people. Findlay and Findlay (1987) predicted that the total world population would increase to 6.8 billion by the year 2000, greatly increasing the demand for food. This is the Malthusian argument, which holds that the rate of population growth will always grow faster than the food supply (Malthus 1798; Lowe and Bowlby 1992). Thus, limits must be set on population growth (Meadows et al. 1972); otherwise, an environmental catastrophe may result, bringing famines, food shortages and political chaos (Ehrlich 1968; Campbell 1998). Some authors reject this thesis, because the pressure of increasing populations on soil and agricultural resources may encourage innovative solutions (Simon 1981). Nevertheless, soil erosion has serious consequences for human life, and the increasing pressure of population on agricultural resources has not improved the situation.

The comparatively rapid population growth rate in the developing world intensifies the utilisation of soil resources (Findlay and Findlay 1987). The combination of improved medical facilities and nutrition levels, healthier life styles, and cultural norms favouring large families has created high population growth rates in Africa, South America and tropical Asia. Between 1973 and 1982, thirty developing countries had population growth rates of over three percent per annum, while the more developed countries frequently had growth rates below one percent (Findlay and Findlay 1987). Soil erosion, in combination with other factors like war and drought, can create famine, as in the Ethiopian and Somalian cases. Between 1960 and 1980, food production in sub-Saharan Africa increased at only 1.7 percent annually, while population grew at 3.2 percent annually (Paulino 1986). A similar situation also existed in the Middle East and North Africa (Lal 1990). Nevertheless, as pointed out by Sage (1994) these environmental problems are not the sole responsibilities of the developing nations. He points out that the industrialised countries present the biggest threat to global resources. This is so despite their comprising only 24 percent of the world's population at the time. Due to the very high consumption levels in the north, the developed nations also put pressure on the agricultural systems in the developing nations to increase production for export.

4.4. Soil Erosion in West Africa

In West Africa as noted above, agricultural intensification and forestry are the most serious causes of erosion problems. As pointed out by Eckholm (1978) land use patterns are manifestations of deep political, economic and cultural structures. Soil erosion is severe in most parts of West Africa, but it is most severe in savanna regions with an annual rainfall of 500 to 1000 millimetres (Lal 1993; Oldman 1994). In the Sahel zone, the low moisture input and sparse vegetation offer little protection to the soil from wind erosion, resulting in soil losses ranging from 10 tons to 200 tons per hectare per year (Lal 1993). In the sub-humid savanna regions with 800 to 1200 millimetres annual rainfall, water is the main agent of erosion. In the forest region, erosion occurs when vegetation is cleared, exposing the soil to the weather (Dickson and Benneh 1988). Mechanised land clearing, with consequent soil compaction and denudation, overgrazing and bush fires are common factors of soil erosion in West Africa. Intensive agricultural systems, the non-enforcement of property rights to prevent over exploitation, inadequate extension agencies, and the lack of public awareness exacerbate the problem.

As pointed out in the section on the soil loss equation, the soil systems of the West African region are particularly susceptible to erosion. The pressure on these fragile soils to support more intensive agricultural systems and growing populations has created severe erosion problems. The management of these problems, and the consequent success or failure of the transition from traditional shifting cultivation methods to more intensive systems, may well determine the future development of West Africa (Coulter 1972, Dickson and Benneh 1988; Elliott and Campbell 2002; Campbell 2005a). Soil erosion may cause many social problems. Water supplies may be affected. For example in Nigeria and Ghana, the siltation of the Kainji and Volta lakes has marginally reduced the irrigation potential of nearby lowlands. Soil erosion on uplands affects water conservation by reducing the onsite storage of soil water in the root zone, creating higher runoff from crusted surfaces, clay removal, and decreases in effective rooting depth (Troeh et al. 1980; Lal 1985; Lal 1993). There may also be a loss of rural livelihoods due to degraded farmland, and excessive rural to urban migration (soil erosion is not the only factor behind this, but it is an important influence) (Foth and Schafer 1980). For example, grain yields on Nigerian alfisols collapsed under soil erosion, with serious social effects (El -Swaify 1993).

4.5. Natural Factors and Effects of Soil Erosion in West Africa.

The air masses, through their impact on temperature and rainfall are important factors for soil erosion. The movements of the Inter-Tropical Convergence Zone (I.T.C.Z.), which divides the dry, northern, tropical continental (cT) air mass from the moist, tropical maritime (mT) air mass, dominate the climate of West Africa (Henderson-Sellers and Robinson 1986). The I.T.C.Z. moves north from April to September, due to the low-pressure system caused by the overhead sun at the Tropic of Cancer. This brings warm, rainy weather from the mT air mass to the areas below approximately 21° N. From December to February, the I.T.C.Z. moves south to about five to 7° N, due to the overhead sun at the Tropic of Capricorn, and the consequent low pressure system in the south Atlantic ocean (Harrison Church 1980). This brings hot, dry, dusty, north-easterly winds from the cT air mass to West Africa. The Equatorial Easterly (E) air mass brings line squalls to the region (Dickson and Benneh 1988). These systems are shown in Figure 2. They are broadly related to the vegetation zones of West Africa, shown in Figure 4.

When the moist, south-westerly winds rise over highland features, orographic rainfall occurs. Examples are the rainy climates of the Fouta Djallon of Guinea, the Guinea highlands of Sierra Leone and Liberia, the Freetown Peninsula, the Mampong Scarp of Ghana, and on the Nigeria - Cameroon borderlands. The relief features may also create rain shadow effects. Temperatures are lowered, and the increased altitude may increase the diurnal range (Nieuwolt 1977; Harrison Church 1980). The cold, southward flowing Canaries current brings cool weather and occasional fogs to coastal Senegal and Cape Verde. The cold Benguela current cools eastern Ghana, Togo and Benin. It also causes dry weather, by restricting convectional conditions (Nieuwolt 1977; Dickson and Benneh 1988).

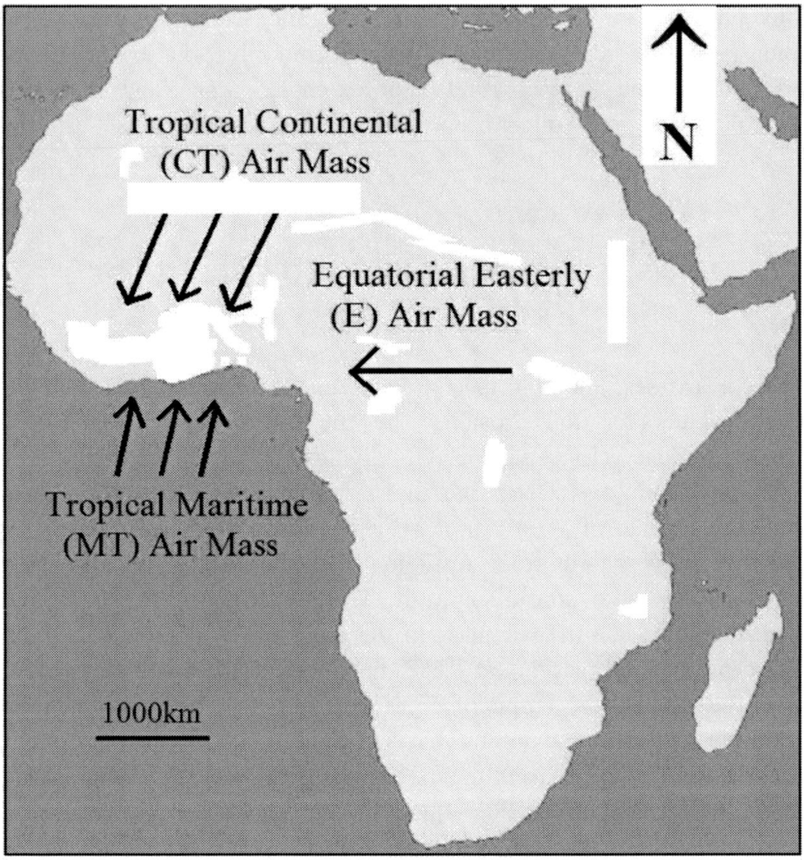

Figure 2. Air Masses of West Africa.

In terms of geology, West Africa is largely composed of folded Pre-Cambrian rocks, (metamorphosed rocks such as mica schists, gneisses, quartzites, conglomerates and amphibolites), with a few sedimentary deposits and Hercynian igneous intrusions such as basalts and granites (Harrison Church 1980; Williams 1984; Dickson and Benneh 1988). Relief has a major effect on erosional drainage patterns. Miocene earth movements (26 million to five million years ago) caused the rivers to cut down to new base levels, increasing the speed and depth of erosion, a situation that still exists today. Examples are the Guinea coast rivers (Stamp and Morgan 1972; Harrison Church 1980). By contrast, the rivers of the sandy, central lowland zones have courses that are more devious, are more mature, and have large deposits of alluvium, due to the gentler slopes and softer ground. These include the upper reaches of the Volta River, the middle Niger and the Senegal. Seasonal flooding is common, especially on the flood plains of the more devious rivers, such as the lower reaches of the Niger. During the rainy season, the river-flooded areas of the Niger between Segou and Timbuktu are as large in area as England and Wales. Tributaries may then become distributaries, as in the case of the Sourou. However, much water is lost due to the intense evaporation and percolation during the dry season (Mitchell 1984; Fischer 1987; Sivakumar and Gnoumou 1987).

The main soil zones of West Africa are (United States Department of Agriculture USDA classification) the Aridisols, Entisols, Alfisols, Oxisols and Ultisols (Figure 3). Under the Food and Agriculture Organisation FAO classification, these Orders are approximately equivalent to Xerosols/ Yermosols, Fluvisols/Arenosols/Regosols, Luvisols/Podzoluvisols, Ferralsols and Acrisols/Nitosols (Nortclif 1988).

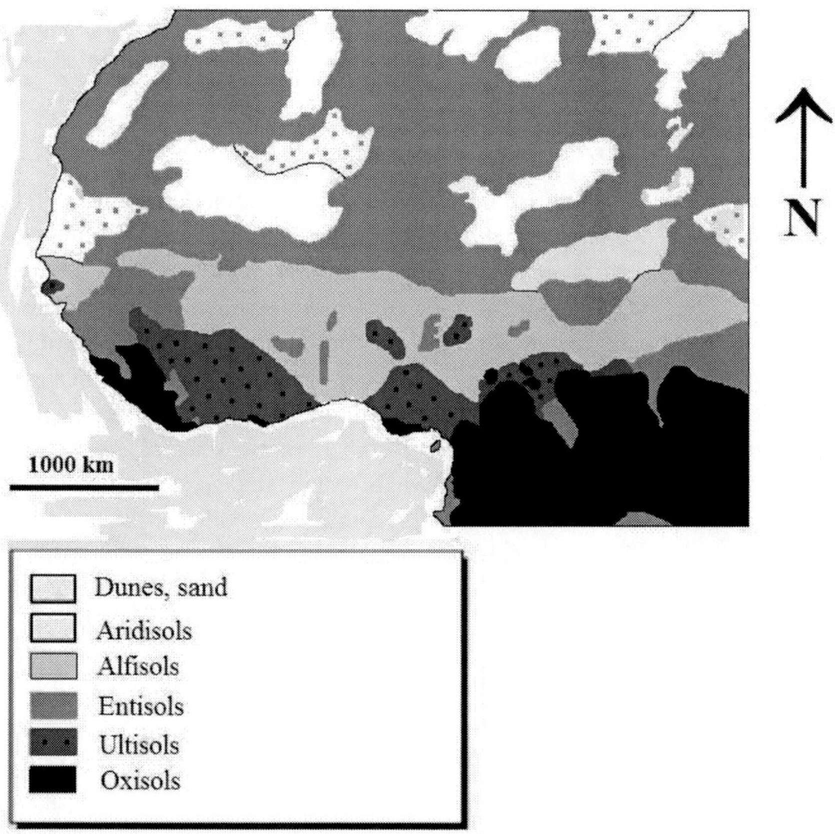

Figure 3. Main Soil Types of West Africa.

Aridisols occupy the driest areas of northern West Africa. They have little profile development, with a light coloured surface horizon that is low in organic matter and may have a calcic, gypsic or saline horizon, or contain some salts. They are highly susceptible to wind erosion and runoff from occasional downpours (Foth and Schafer 1980; Brady 1984; Mitchell 1984). Entisols are soils of recent origin, mostly derived from alluvial materials, with usually no horizons except an A-horizon. Entisols are very diverse, the most extensive of the soil orders, because they generally include most soils that do not fall into any of the other soil orders. In West Africa, they are generally found with aridisols in the Sahara desert and the Sahel Savanna, and are prone to erosion. Alfisols, the main soils of the West African savanna, are derived from crystalline, acid rocks, with a high quartz content and a low iron content. They are usually characterised by an argillic B horizon, a base saturation of more than 35 percent, a gravelly horizon at shallow depths and a light textured surface horizon (although some variations have a coarse textured surface horizon), and a high proportion of kaolinitic clay (Foth and Schafer 1980; Brady 1984; Ahn 1993). There is frequently bedrock within 100 centimetres of the surface. They are highly prone to compaction and erosion by water (Lal 1993).

The oxisols and ultisols are the highly leached, drained soils of the forest zone. Oxisols are the most highly weathered soils in West Africa, but their occurrence in this area is limited. Their main area of distribution is Central Africa. They are the predominant soils of the equatorial rain forest climatic zone (Foth and Schafer 1980; Ahn 1993). There is intensive and continuous leaching and weathering, creating very deep profiles, loss of silicates by hydrolysis and accumulation of iron and aluminium. The high temperatures and moisture levels, and the consequent high biological activity, create very high rates of organic decomposition, hampering the accumulation of soil organic matter. As the soil pH declines, aluminium and iron become more soluble, reacting with the phosphorus to make it unavailable to plants. With further declines in the soil pH, aluminium toxicity may limit plant growth and reduce biological nitrogen fixation (Foth and Schafer 1980; Brady 1984; Dickson and Benneh 1988; Ahn 1993; Lal 1993). Ultisols are also highly weathered soils, but less so than the oxisols. The base saturation of the argillic B-horizon is lower than that of the alfisols (below 35 percent). Due to the low cation exchange capacity of the dominant mineral (kaolinite), the fertility of these soils is low. Variations in these soils largely reflect the length of the dry season. Under heavier rainfall, calcium, magnesium and potassium are readily leached out of the profile. Exchangeable aluminium accumulates and may be toxic to some plants. Iron oxides may also accumulate, giving the soil a red, brown or yellow colour, depending on the degree of hydration. The subsoil has poor permeability, and is prone to erosion and waterlogging (Foth and Schafer 1980; Brady 1984; Ahn 1993; Lal 1993).

In terms of vegetation, the main types of vegetation in West Africa in order of latitude are; the semi-Sahara and Saharan vegetation, Sahel, Sudan and Guinea Savannas, Dry Deciduous Forest, Moist Deciduous Forest and Rain Forest.

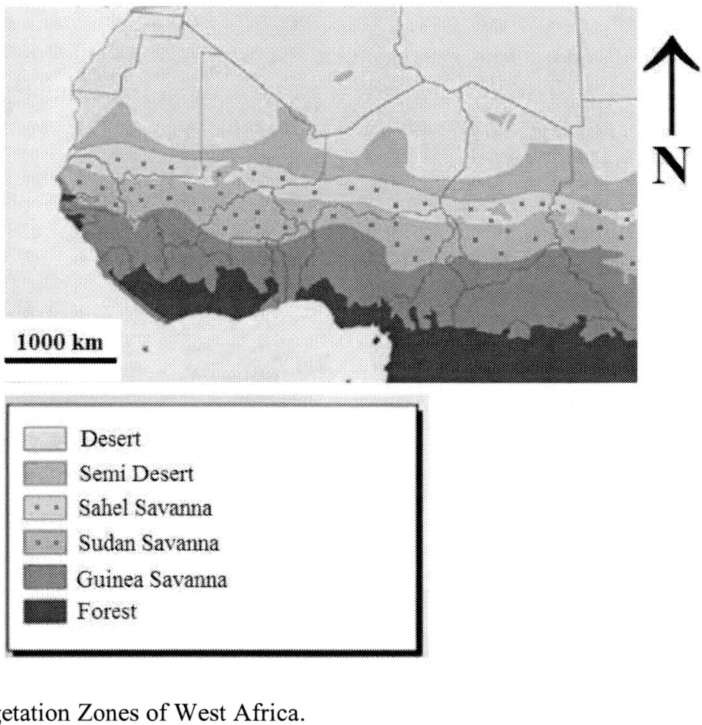

Figure 4. Vegetation Zones of West Africa.

The semi-Saharan and Saharan vegetation comprises dispersed grasses, annuals and very rare trees. The occasional rain allows seasonal herbs to spring up (Harrison Church 1980). The Sahel Savanna has a dry season of at least 8 months. There are a few trees, with spreading roots that hold the soil together. Common species are *Acacia seyal* (which prefers clay depressions), *Acacia tortilis raddiana* (which prefers light sandy soils), *Acacia senegal*, *Acacia laeta*, *Acacia ehrenbergiana* and *Acacia nilotica*, and thorn shrubs such as *Commiphora aficana*, *Balanites aegyptiaca*, *Euphorbia balsamifera*. Short grasses are also common (Harrison Church 1980; Dickson and Benneh

1988). The Sudan Savanna is more moist, with a long dry season of almost 7 months. The few tree species include *Hyphaene thebacia* (dun palm) *Sclerocarya birrea*, *Balanites aegyptica*, *Adansonia digitata* (baobab), *Tamerindus indica*, *Ceiba pentandra* (kapok) and *Acacia albida* (Harrison Church 1980; Dickson and Benneh 1988). The Guinea Savanna has a higher rainfall regime. It is characterised by grass 1.5 to 3 metres in height, and broad leaved deciduous trees mostly 12 to 15 metres high, occurring in clumps. Outliers of lowland rainforest occur on moist, well-drained slopes, and along rivers. Common trees are *Anogeissus leiocarpus, Afzelia africana*, *Detarium microcarpum*, *Vitex doniana*, *Hymenocardia acida*, *Daniella oliveri*, and *Terminalia glaucescens*. Grasses include species of *Andropogon*, *Hyparrhenia* and *Pennisetum*. In the drier northern areas, burning is a common feature, but where it is not very prevalent the dominant species are *Isoberlina doka*, *Isoberlina dalzieii*, *Monotes kerstingii* and *Uapaca togoensis*. Trees of economic value include *Butyospermum parodoxum* (shea butter tree) and *Tamarindus indica* (tamerind).

Another type of savanna vegetation is the Ghana coastal scrub and grassland. Here, centuries of human activity, combined with low rainfall, and possibly cool winds from the cold offshore Benguela current, have created a zone of dense scrub and grassland on the south east coast of Ghana, commonly termed the Accra plains, isolated on the otherwise forested West African coast (Dickson and Benneh 1988). There is also the Senegal coastal vegetation, of coastal climatic zone of Senegal, which has richer vegetation than the Sudan savanna zone, due to the greater atmospheric humidity and the higher water table. In clayey areas, baobab *Adansonia digitata* and the *Borassus aethiopum* palm are common (Harrison Church 1980).

The lowland rainforest, also termed the closed forest, is divided into evergreen rain forest, moist semi-deciduous forest, and dry-deciduous forest (Dickson and Benneh 1988). The evergreen rainforest, with over 1900 millimetres of rainfall annually, is characterised by a triple layer of canopies, an undergrowth of epiphytes and woody climbers, rapid decomposition of the leaf litter layer due to the high temperatures and moisture facilitating intensive microbial activity, continuous plant growth and the absence of a dry season. Further north, the lower rainfall (usually between 1250 and 1750 millimetres) and the presence of a dry season (usually from November to March) create the moist and dry semi-deciduous forest, which is less rich in vegetation (Harrison Church 1980; Dickson and Benneh 1988). Common timber species are *Khaya ivorensis* and *Khaya anthotheca* (mahogany), *Entandrophragma cylindicum* (sapele), *Guarea cedrata* (guarea), *Tieghemella heckelii* (makore), *Lovoa trichilioides* (African walnut), *Pipteniastrum africanum* (Dahoma or Ekhimi) and *Triplochiton scleroxylon* (obeche, wawa or samba) (Harrison Church 1980; Dickson and Benneh 1988). There is also Casamance woodland, limited in extent to the Casamance district of Senegal, Guinea Bissau and lowland Guinea, which may be distinguished from the lowland rainforest by its long dry season, poorer species composition (mainly *Parinari excelsea*, *Erythroleum guineense*, *Afzelia africana* and *Detarium senegalense*), low branching and spreading trees, and a thicker layer of undergrowth composed of bushes, lianes and herbaceous plants (Harrison Church 1980).

The modified lowland rainforest may be classified, depending on the degree of modification, into: (1) depleted forests, composed of a patchwork of primary and secondary growth, which results from selective timber felling; (2) secondary forests, which are the regenerated growth after shifting cultivation. In the early stages, these comprise shade intolerant species such as *Musanga cercropiodes* (umbrella or pastoral tree), *Chlorophora excelsa*, (Iroko), *Terminalia superba* (Afara) and *Triplochiton scleroxylon* (samba), with forest growth occurring later; and (3) derived savanna which occurs when grasses invade cleared areas, annual fires (man-made or otherwise) kill off most of the remaining trees, and fire-tolerant trees, such as *Lophira lanceolata* (red ironwood), colonise the grasslands, as in the Sobo plains in southern Benin (Harrison Church 1980; Dickson and Benneh 1988).

Other types of forested vegetation include montane vegetation. This, affected by increased and more constant humidity, lower temperatures and evaporation rates, includes rain forest (with bryophytes, tree ferns lichens and epiphytes), drier, grassy woodland,(as in the Bamenda highlands) and savanna derived from the original climax vegetation of montane woodland (common in the Loma mountains of Sierra leone) (Fenley 1979; Harrison Church 1980; Dickson and Benneh 1988). There is also mangrove forest, composed mainly of mangrove trees, which correlates with the southwest coast monsoonal, equatorial and Liberian climatic regions, and occurs near lagoons and deltas. Common species are *Rhizophora racemosa*, *Rhizophora harrisonii*, and *Rhizophora* mangle. The stilt roots of mangroves may encourage the deposition of mud in coastal areas, and moderate the erosive effect of tidal waters (Harrison Church 1980; Sayer et al.1992). There is also Fresh Water Swamp Forest. This vegetation is characterised floating grass or sudd, shrubs and stilt rooted trees, open forest canopies and climbing palms. It is common in the Niger Delta (Sayer, Harcourt and Collins 1992). Common species are *Mitragyna ciliata*, *Pandanus candelabrum*, *Rhaphia hookeri*, and *Raphia vinfera*. (Harrison Church 1980).

In terms of the natural effects of soil erosion, firstly, there are onsite effects of soil erosion. The relationship of soil erosion to soil fertility and agricultural productivity in the West African savanna is very close. Nevertheless, the impact of erosion tends to be higher on highly weathered soils such as the oxisols and ultisols in forest clearings, because their initial fertility is low and most of the soil's fertility occurs in the topsoil due to the deeply drained nature of the subsoil. Rooting depth in such soils also tends to be limited were extensive erosion of the surface has occurred. This may impede the efficiency of water use by plants. Crop sensitivity is greatest for cereals, followed by legumes and root crops (El-Swaify 1993).

Secondly, there are offsite impacts of soil erosion. These occur mainly downstream from erosion prone areas, such as where increased human populations occupy the upper reaches of river basins and watersheds. Excessive water and sediment losses combine to produce the siltation of water bodies, and the deposition of sediments on low-lying farmlands, roads and other structures (El-Swaify 1993). Such problems are common in the southern river plains.

Thirdly, there is desertification. The removal of the topsoil, following vegetation clearing, degrades the soil's fertility and leaves bare rocks and weathered, sandy and stony subsoils on the surface, sometimes permanently hampering vegetation recovery. Desertification is a serious problem in the Sahel and Sudan savanna regions and in some areas, the desert is encroaching on towns and villages (Derrick 1984; Milas 1984; Middleton 1985); Brittain 1994). Desertification of lowland areas may force farmers to cultivate and deforest mountainous slopes, with serious erosion consequences (Packham et al. 1992).

4.6. Social Factors and Effects of Soil Erosion in West Africa

One factor for soil erosion is agriculture. Traditional agricultural practices in West Africa are shifting cultivation, bush fallowing and permanent plots (mainly in the forest region), and cattle and goat herding (mainly in the Sahel, Sudan and Guinea savanna zones). The traditional shifting cultivation system combines a period of fallow (usually about 8 to 20 years) with a period of cultivation (usually about 3 to 5 years) (Foth and Schafer 1980; Dickson and Benneh 1988). Under shifting cultivation, a good equilibrium between humans and their environment may result (Reijntes et al. 1992). Nevertheless, the actual environmental effect of shifting cultivation is debatable. This system is widely blamed for the removal of vegetation, soil exhaustion and erosion (Barrow 1991). Traditional African cultivators are sometimes portrayed as wasteful or destructive of tropical forests due to their lack of environmental awareness, ignorance or apathy (Blaikie (1985). The criticisms of traditional farming practices are usually based on environmental managerialism, which blames environmentally incompetent cultures, population pressure, and social ignorance for the incidence of soil erosion. There is seen to be an optimum balance of natural resource use for the combination of productivity with sustainability, the traditional farmer being a deviant from this line (Redclift and Woodgate 1993).

Baker (1981) and Blaikie (1985) challenge this managerial approach, by pointing out that it does not fully explore the historical, economic and political background. The emphasis is on symptoms rather than causes, and it blames people for actions arising out inequality. An alternative paradigm they advocate is the political economy (or political ecology) approach, in which environmental degradation is seen as a manifestation of political and socio-economic inequalities, such as unequal access to land, the marginalisation of the poor and the lack of alternatives for those short of land. Sometimes the poor may be forced to destroy their own environment in their attempts to delay their own destruction (Gallopin and Berrera 1979). These are symptoms of the problem, which is the political economy of the country (Allen 1980; Baker 1981).

The political economy approach to the problems posed by shifting cultivation hints that while under the traditional cultural context the people accurately perceived the need for extended fallow periods to restore the fertility of fragile tropical soils, the current international economic system dictates the development of extensive areas of cash crops for export, marginalising the small scale farmers who produce food crops, and forcing them to either change to more intensive systems, or change to a new lifestyle. The pressure of increasing populations on unequal distribution of land increases the problem. Therefore, the effect of shifting agriculture on the soil depends on the total environment in which it is practised.

Inequality of land ownership is a major factor behind migratory movements in West Africa, which increased during the 1980s (Pieri 1992). The large-scale intrusion of cash cropping systems, dictated by the international economic and political system, caused people to move to areas of intensive agriculture to find jobs. A problem emerges when migrants receive inferior land for subsistence to supplement their meagre income. In northern Cameroon, the Muslim Foulhe only allowed the migrant Guiziga to settle on low quality land, encouraging erosion and degradation (Pieri 1982). Other examples are groundnut cultivation in Mali and Niger and of pure stand maize and cotton cultivation in the deciduous forest and guinea savanna. Belem (1985) and Tallet (1985) noted that while the population growth rate for Burkina Faso during the early 1980s was 2.4 percent per annum, the growth rate for the cotton zone, which was expanded to increase export earnings and provide jobs, was over 3 percent per annum. In Mali, the peasants were losing the landuse struggle against more powerful, cash cropping interests, cultivating instead marginal, fragile soils (Blaikie 1985). However, the government blamed the peasantry for sheet and gulley erosion (Lipton 1982).

Other factors such as inappropriate foreign technologies, political manipulation of the production system and economic priorities may combine to create environmental problems. Mechanised clearing of trees, stumps and roots, and ploughing increases soil erosion (Crasswell 1993). The extent of soil erosion depends on the water content, soil depth and time of rainfall in relation to the tillage operations, but the erosive effects are higher in sandy soils (such as those under Casamance woodland) than in clayey soils (such as vertisols) (Fauck 1977).

Cattle herding, also an important traditional agricultural activity, takes place mainly in the Sahel and Sudan savanna zones, due to the occurrence of the tsetse fly (the vector of the disease trypanosomiasis or sleeping sickness that affects cattle) in the forest zones. Nomadic cattle herding may result in soil compaction and the consumption of scarce vegetation, especially along the migration routes to the south. During droughts, such as those that occurred through most of the 1970s and 1980s, the herdsmen were forced to graze their cattle over large areas to find forage, increasing the effects on the soil. Similar problems occur for sheep and goats (Harrison Church 1980).

Another factor for soil erosion is population growth. It has been argued that any attempt to deal with environmental deterioration must first deal with the basic cause: overpopulation (F.A.O. 1980). West Africa, with only about 20 percent of the total area of Africa, has nearly one third of total population of Africa, which is estimated at just over one billion (Harrison Church 1980; Central Intelligence Agency 2012) (Figure 5). This shows the fast increase in the population over the period 1985 to 2012. The population distribution of West Africa is due largely to historical factors. Trading posts with trans-Saharan Arab traders and European maritime powers attracted settlers. Examples are the port of Takoradi in Ghana, and the towns of Gao and

Timbuktu in Mali. Traditional land attachment explains the high populations of the Dogon area of Mali, the Mossi area in Burkina Faso and the Igbo area of Nigeria.

Country	Area	Population (1985-90)	Population (2012)
Burkina Faso.	274,200.	8,776, 000	17,275,115
Cameroon.	475,442.	11,554, 000.	20,129,878
Gambia.	11,295.	848, 000	1,840,454
Ghana.	238,537.	14,425, 000	25,241,998
Guinea.	254,857.	5,547, 000	10,884,958
Guinea-Bissau.	36,125.	960, 000	1,628,603
Cote d'Ivoire	322,463.	11,713, 000	21,952,093
Liberia.	111,379.	2,475, 000	3,887,886
Mali.	1,240,000.	8,212, 000	14,533,511
Mauritania	1,030,700.	1,954, 000	3,359,185
Niger.	1,267,000.	7,479, 000.	17,078,839
Nigeria.	923,768.	113,665, 000.	170,123,740
Senegal.	196,192.	7,211,000	12,969,606
Sierra Leone.	71,740.	4,040, 000.	5,485,998
Total	6,453,698	198,859,000	326,391,864

Figure 5. Demographic Statistics of West Africa. Source: World Bank Atlas (1990) in Binns (1994), Central Intelligence Agency World Fact book (2012).

Defence considerations also played a role, such the Yoruba war camp that developed into Ibadan, Nigeria. The political stability of past Hausa, Fulani and British governments encouraged much settlement in Nigeria. Religious settlements also developed, especially in Muslim areas. Mining settlements developed, for example, Bukuru in the Nigerian tin fields, and the gold mining towns of Tarkwa, Prestea and Obuasi in Ghana (Harrison Church 1980; Dickson and Benneh 1988). Modern commerce and industry expanded towns such as Dakar (Senegal), Abidjan (Cote de Ivoire), Tema (Ghana), Lagos and Port Harcourt (Nigeria) (O'Connor 1991; Binns 1994). Soil fertility and rainfall, and consequent agricultural potential, also encouraged settlements in the southern third of West Africa. Population concentrations created increased pressure on the soil through agricultural intensification. These, however, were negatively affected by the occurrence of disease such as malaria, caused by microbes of the genus *Plasmodium* and spread by the *Anopheles* mosquito; and sleeping sickness or *Trypanosomiasis*, caused by the microbe *Trypanomsoma,* and spread by the fly *Glossina palpalis (*Stamp and Morgan 1972).

The irregular distribution of the population of West Africa creates problems. The situation in northern Togo is a typical example. In the more heavily populated areas the land is impoverished, while in the rest only about 60 percent of arable land is utilised (Raymond and Tourte 1985). In areas of high concentration, the high prevalence of poverty compels the people to utilise fuelwood as the main source of energy. Anything combustible is burnt, diverting crop residues from fertiliser to fuel uses, and hampering the battle against soil erosion (Agarawal 1980; Bailly et al. 1982; Pieri 1992). A similar problem exists on the Mossi plateau at Yatenga (Marchal 1983) in southern Mali (Pieri 1992), and the areas around Ouagadougou in Burkina Faso, which are treeless with a soil that is bleached, impoverished in organic matter, structureless, eroded and largely abandoned (Wane Conde 1985).

The intensive population increase in the savanna areas of West Africa, coupled with the change from subsistence to a market economy has influenced the extensification and intensification of agricultural systems. Sometimes the traditional system does not change markedly, but expands. In the Oudalan region north of Burkina Faso, where the population increased by 64 percent during the 1980s, the cropped area was increased by a reduction and/or abolition of the fallow period, and the extension of farming onto less fertile areas (sand dunes, the lower slopes of hillocks and clayey low-lying areas). Soil deterioration was marked by gulley erosion, and stunted millet crops resulting from nitrogen deficiency. Thus, intensification with no change in technique can cause problems (Milleville (1980). A second method is to change the method with intensification. Pelissier (1966) noted that the Serer of Senegal developed an intensified system to support the concentrated population, soil fertility and the environment. Cattle and *Acacia albida* shared the land between cropping and fallow periods. Cereals (millets) and legumes (groundnuts) were rotated during the cropping periods. The minimal soil erosion that resulted shows that population increase does not necessarily cause soil erosion, but the combination of favourable political, economic and social factors can ensure an environmentally favourable result (Pieri 1992).

In the Maradi district of Niger, due to increased population pressure, the fallow period was reduced, but regular manuring and the incorporation of crop residues kept soil erosion to a minimum (de Miranda 1977; Raynault 1980). Other traditional cultures have prevented the denudation of the landscape. In the region around Kano in Nigeria, the areas with the highest rural population densities (up to 550 persons per square kilometre), close to the urban woodfuel market, had the highest tree densities (12.3), and even recorded an annual increase (between 1972 and 1981) in tree numbers of 2.4 percent, due to extensive planting. Other areas further away from the city also recorded a 26 percent increase in tree density between 1972 and 1985 (Binns 1994).

Thus, population growth is a contributory factor rather than a direct cause of soil erosion. Some even believe that population growth and the resultant pressure on resources creates the incentive for developing better resource systems (Simon 1981). Examples may be found to illustrate these opposing views, but an examination of the context must conducted before a judgement of the relationship between population and soil erosion may be made.

Forestry is an important factor for environmental change (Elliott and Campbell 2002). As argued by Harrison Church (1980) the survival of rain forests beyond another century seems doubtful. About 1.3 million hectares of closed broadleaved forests in Africa were cleared annually during the 1980s, and 2.3 million hectares of open woodland were lost each year (Dankelman and Davidson 1987). More than half of these losses took place in West Africa. The primary forest in Sierra Leone was reduced from 60 percent of the country's area to 6 percent, while in Cote de Ivoire the figures were a reduction from 38 percent to 8 percent (Brittain 1994). The West African savanna is possibly derived from exploited forest. Forestry however is not the only cause; clearing for agricultural landuse and firewood is a complementary problem. However, many scholars have argued that not enough is known for such an assumption and the factors for vegetation change, namely fire, herbivory, nutrients, moisture and human factors are too dynamic. Therefore, savanna may be preceded by forest, but there may be continual change, due to the interaction of numerous natural and human factors in dynamic coevolution (Solbrig 1993; Campbell 1998).

Urbanisation is another important factor for soil erosion. Soil erosion results when land is cleared for settlements and/or transportation networks. There is massive rural to urban migration throughout West Africa, expanding settlements in the suburban areas, and invading areas that may be termed green belt zones (O'Connor 1991). At least a third of Africa's urban population live in slums or squatter settlements, and these are growing by at least 15 percent per annum (Binns 1994). The environment in these overcrowded and dilapidated settlements encourages erosion because most of them are built on bare ground. In some cases, the dumping of wastes toxic to plant growth destroys empty lots otherwise occupied with vegetation. Little or no vegetation grows in such settlements, even in garden cities like Kumasi, Ghana. Exposed areas include pot-holed roads, paths, sidewalks composed of compacted earth, open gutters, poorly drained low-lying areas and market centres in central business districts. The problem increases by the roads and pathways, which combine concrete with compacted soil, encouraging high runoff. Where land is available, farming may take place at the urban fringe, creating serious erosion problems, especially in areas of intensive competition. Urban dwellers, (mostly women), may plant crops to increase the household food supply, while others (mostly men) grow crops for commercial purposes. In Kano, Nigeria, there is competition for land between rural and urban dwellers, especially over fadama land where the water table is close to the surface, and irrigated vegetable production is possible throughout the dry season. Other examples are the mountain villages around Freetown in Sierra Leone, and the horticultural areas around Banjul (Binns 1994).

In terms of the social effects of soil erosion, the resultant lowering of soil fertility, loss of agricultural land and decreased food production may give rise to famine (Pieri 1992). While famine in West Africa was not as disastrous as in Somalia, Sudan or Ethiopia, it is still serious, because even normal food production levels fall well below what the World Health Organisation regards as the minimum necessary for health. There is widespread malnutrition, contributing to diseases kwashiorkor and similar ailments. Seasonal hunger was common, especially in the savanna areas (O'Connor 1991). Widespread famines occurred for the first time in 1973-74 (largely because of the prolonged droughts of the late 1960s and early 1970s, which caused high levels of wind erosion), and again in 1985 (after the drought of 1983-84). It is arguable that the droughts were a consequence of agricultural intensification, with the consequent soil erosion reducing the vegetation cover and the soil moisture, exposing the area to the dry Saharan winds and desiccating the landscape. During the 1980s famine, harvests deteriorated, and at least one million people were negatively affected in the Sahel and Sudan savanna zones. Mass starvation was avoided because more than a quarter of a million tons of grain was shipped from overseas to each of the affected countries (O'Connor 1991).

While soil erosion, drought and even civil war were blamed for the occurrence of famine, Sen (1981) argues that poverty and inequality in access to food are the basic underlying socio-political contributory factors behind the occurrence of famine. Nevertheless, while proportionately more people starve, there may be a real reduction in the food supply, causing a social dislocation at all levels of the socially. Thus, the problem was due to both a lack of food and a lack of access to it; it is a political economic, social and environmental problem (O'Connor 1991). Some nations, such as Nigeria and Ghana, saw their food import bills rise significantly during the 1980s. This cut down on the expenditure on social services. For example, the drought of 1983-84 in Ghana caused a lowering of the level of social services, bringing in its wake public discontent (mainly by university students and the military) which nearly brought down the government. Another consequence of the degradation of rural livelihoods was the increase in rural to urban migration. This is especially true in the Sahel regions of Mauritania, Niger, Mali, Burkina and Senegal (Dankelman and Davidson 1987; Binns 1994; Brittain 1994).

CONCLUSION

This chapter has examined the natural and social issues of soil erosion in West Africa. The dominant issue concerned the complexity of the problem from both natural and social perspectives. Comparison of regional scenarios is difficult, as not only do soil forming factors and social structures vary, but isolated unpredictable events such as droughts occur, creating further scenarios of conflict, impoverishment, development and adaptation. One method for developing further understanding of such a complex problem is to focus on a case study. Such a case study is not necessarily typical for the West African region. Rather, it hoped that deeper understanding of the issues might emerge from a detail of assessment, with some commonality with other areas. The next chapter will look a Ghana as a case study.

Chapter 5

Soil Erosion in Ghana

5.0. Introduction

This chapter examines the issues presented in the first four chapters, in application to Ghana. Ghana is a fair example of the complexity of the relationship between development and environmental issues. Agricultural development and environmental change in Ghana rest on the key issues of traditional agricultural systems, the colonial introduction of cash cropping, the competing development models (centralised state planning and market economy models) in the independence period, the demographic transition and agricultural intensification, and the resultant environmental problems of soil erosion, deforestation, desertification risk and periodic drought. However, this historical chronology of agricultural development is complicated by the heterogeneous, multicultural nature of the nation, the consequence of the integration of several African tribal states into the British demarcated Gold Coast. Traditional socio-cultural and socio-economic systems must thus be appraised with the political side of political ecology and social theory. Many ethnic groups inhabit varying environments, in particular forest and savanna, and this has shaped their particular environmental outlooks and agricultural systems, and influenced their reaction to the introduced cash cropping export-oriented systems. The complexity of these variables underscores the need for an integrated analysis.

5.1. Agricultural Development, Soil Erosion, and Social Effects

In Ghana (population 24,233,431, area 238,540 km^2 CIA 2012), the dominant landscape is of mostly flat, forest and savanna vegetation, with more rainfall in the forested south than in the savanna covered north, with the exception of the dry coastal savanna (Figure 3, Figure 6). The variability of the northern savannas and southern forests affect land use (Sarpong and Asuming-Brempong 2004). For example, the northern tribes live in the Guinea savanna, relying mainly on cattle herding and cereal crops (with desertification risk being the main environmental problem), while the Akan of the deciduous forest combine the traditional food cropping of root crops with cocoa and coffee cash cropping and forestry (deforestation is the main environmental problem). Mountainous areas cover less than 15 percent of the total land area (Dickson and Benneh 1995). Although gentle topography was historically prized for settlement and agriculture, the study of this issue is relatively neglected in the literature (Taylor 1952; Benneh and Agyepong 1990; Elliott and Campbell 2002). As a result, few studies documented the landuse conflicts over preferences for gentle topography and the biogeographical impacts of this dynamic (Alo and Pontius 2004; Mikkelsen and Langohr 2004; FAO 2005).

The traditional agricultural systems involved local, mostly gender neutral food farming (plantains, cocoyam, bananas, yam and cassava in the forests; millet, grain crops, guinea corn, maize and groundnuts in the savanna) (Pearce 1992). The main methods of cultivation were bush fallowing (a form of shifting cultivation involving temporary plots around permanent settlements) and permanent systems (involving intensive fertilisation-manuring, household refuse and fowl droppings- of permanent plots around compound settlements, and livestock such as goats, sheep and chickens). The socio-political factors behind the evolution of agricultural systems relate mainly to state structure, land ownership and tenure arrangements, class structure, and land access (Awoonor 1990). Land tenure was largely based on small family owned plots and larger holdings owned by traditional elites. Stakeholders were mainly local people engaged in multiple activity livelihoods, including farming, fishing, hunting and trading. The agricultural system was dominated by the structural influences of values (religious and authority structures), environmental perception and knowledge, locally derived technology, and social organisation. Power relations were based on age, gender, inheritance and traditional authority, these in some cases affecting land and other resource access (Campbell 1998).

In the organisation of the agricultural land tenure systems, social values were manifested when individual users of the land were obliged to contribute agricultural produce towards communal occasions such as festivals and funerals. In addition, within allocated plots, some areas were still the preserve of the community. For example, among some of the tribes of south-central Ghana, big trees, rivers and other water bodies were not necessarily included in the user rights bestowed in any land allocation (Awoonor 1990). The "land holding corporation" had two entities: the family, represented by an elder, and more generally, the community headed by the chief (*Ohene*). Over these was the state, headed by the Paramount chief or *Omanhene*. In the fringes of tribal land, the *adikrofo* or village headmen were the main agents and controllers of land allocation (Dunn and Robertson 1973: 225). Strangers occupying tribal land were obliged to make certain payments to the local chief (members of the tribe were exempted). Among the Asante, such a system was termed the abusua system: strangers granted such land were obliged to give one third of their produce to the local chief (Howard 1978).

Insufficient evidence exists for a conclusive judgement on the environmental sustainability of pre-colonial agricultural systems. It may, however, be argued that the traditional system was sufficient for the needs of the particular context at that time, and as they exist today such systems must be sustainable even in the face of socio-economic change. The environmental effects of pre-colonial agriculture related mainly to the impact of shifting agriculture on the natural vegetation, in terms of deforestation and savanna encroachment in areas of population concentration. In some cases, these effects were exaggerations (Fairhead and Leach 1998).

The introduction of cash cropping systems, and associated political, economic, socio-cultural and environmental structures tied to the international, mainly British marketing system began in the second half of the nineteenth century. Kimble (1963:128) noted that "economic development was the main mediating force whereby the individualist, competitive, acquisitive attitudes and values of the West were introduced into African society." The colony was divided into provinces, each with agricultural and mining developments. New powerful stakeholders included British expatriates, local intermediaries, plantation and mine owners and colonial demonstrators. The modern, industrial, rational perception appraised the environment as a tool which, when properly modified and developed could provide an economic justification for the maintenance of the colony. Knowledge was based on technology transfer and scientific universalism, rather than on long-term local experimentation and religious values. Scientific developments led to the importation of agricultural technology, and while this increased agricultural commodity production, it led to an increase in environmental degradation (Richards 1985).

Cocoa *Theobroma cacao* L. introduced into Ghana in 1858, was the main cash crop. Other cash crops were rubber *Hevea brasilensis* and oil palm *Elaeis guineensis*, but these declined soon after 1915 as local farmers and the colonial government preferred the stronger demand for cocoa (Howard 1978). Forestry also developed. Additionally, cultivated crops included coffee, bananas, tobacco, coconuts, sugar and jute (Dickson and Benneh 1988). There were serious environmental effects due to the resultant marginalisation of local structures, the intensification of food production for new markets, the reduction of fallow periods due to the reduced area available, and the increase in demand for food that came with the increase in population during the post independence era. The introduction of inappropriate machinery (such as tractors, which compact the soil), and the increase in forestry and fuelwood extraction, greatly reduced the forest vegetation and degraded many soils (Milner 1989). For example, some of the rice growing areas of northern Ghana were destroyed because of problems with soil fertility and soil erosion (Benneh 1985).

The traditional agricultural and cash cropping systems eventually existed side by side in numerous rural contexts, and in many cases individual rural actors adapted by participating in both systems. People would work as cocoa harvesters, gold miners or construction workers, and maintain the traditional farms, hunting, fishing and/or trading activities. Thus, the dual economy developed, where cash cropping was an aspect of the multiple activity livelihoods. The forms of participation would be classified by Pretty (1995) as actually non-participation, as the local actors worked as automatons, and did not set the agenda or derive autonomy (Campbell 1998). To state that broad structural change occurred in the country is not entirely accurate; what obtained was a myriad of actor decisions that ensured that an observer would see cash cropping and industrialisation and look at the same context and see traditional livelihoods and different forms of environmental change in a complex milieu. Overall, however, there is evidence of gradual deforestation and drier climate, the result of increased land utilisation with population increase, and less appropriate landuse techniques (Campbell 2005a).

In terms of the environmental effects, a common issue in the colonial literature is the increase of deforestation with agriculture and mining development, and disputes as to the responsible group; the local people or the colonial authorities (Howard 1978). Howard (1978:45) noted, "The chief issue of debate between Africans and Europeans was deforestation, which in the cocoa areas was having the effect of reducing humidity". Local farmers blamed European owned mines as being destructive of forest vegetation, while Europeans blamed shifting cultivation. Manshard (1974: 181) blamed cocoa farming and improved communications for deforestation, and cast a bleak picture for the future when he stated, "there will not be any unprotected forests left in the near future". Ormsby-Gore, a visitor to the Gold Coast in 1926, also blamed cocoa cultivation when he argued (in Kay 1972: 206): that the "real" problems of the cocoa industry were the depleted soils and deforestation.

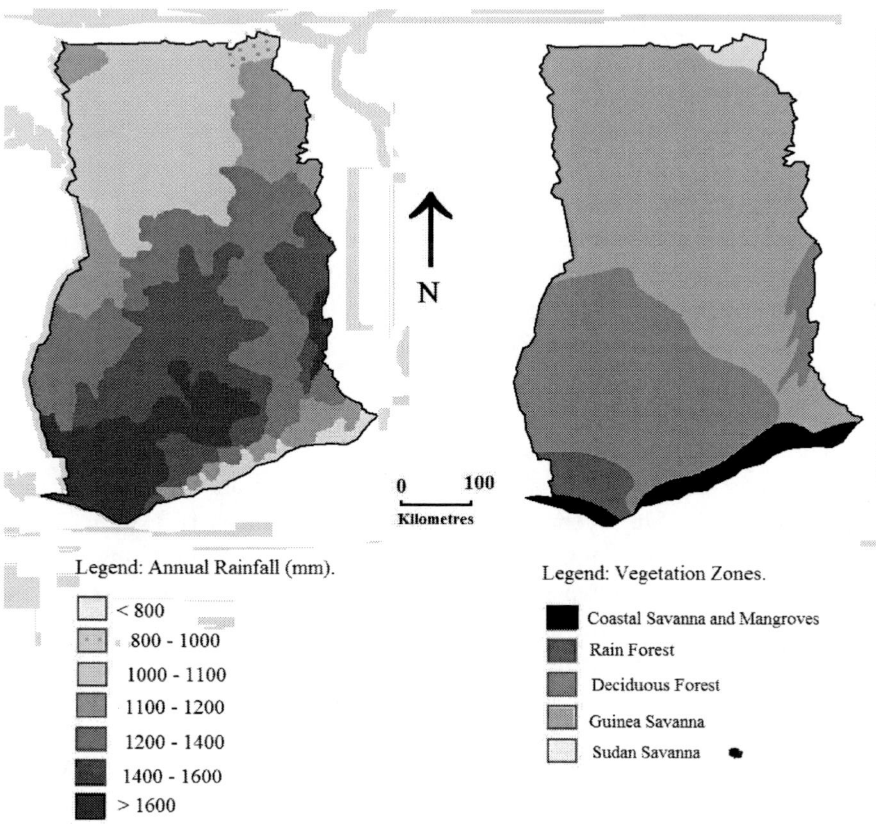

Figure 6. Rainfall and Vegetation Zones of Ghana.

In general the documented trend was of deforestation and soil erosion, but there were disputes as to the causes (mining, shifting cultivation and cocoa farming). Today the issue of deforestation is still very important (Agbesinyale 1992; Gyasi *et al* 1995; Hawthorne and Abu-Juam 1995; Satti 1995). The loss of forest stands in Ghana in the last 100 years is argued to be one of the highest in Africa (Dixon, et al. 1996; Palo et al. 1996), but some authors also examine 'positive' issues such as reforestation and forest management (Parren and Graaf 1995; Richards 1995). Fairhead and Leach (1998) stress the importance of a more critical look at the issue of deforestation as a historical phenomenon, noting that in many cases there was insufficient evidence, poor judgement and generalisation of contextual findings.

The incidence of droughts in 1970, 1975, 1977 and 1983/84 increased soil erosion, affecting farmlands and settlements (Dickson and Benneh 1988; Pearce 1992). Bush fires, both intentional and unintentional, also increased during the dry period (November to May), and became a very important cause of vegetation destruction (Benneh and Agyepong 1990). In the twentieth century, the effects of soil erosion included the substitution of erosion resistant crops (such as cassava) for more sensitive crops (such as plantain and cocoyam), more labour and material inputs to sustain a given level of output, landlessness or unavailability of good farming land close to settlements, declining water supplies and agricultural drought, food shortages in the dry season and famine in the drought years, and emigration of people to urban or more fertile areas. The net effect was poverty and frustration (Benneh and Agyepong 1990)

In northern Ghana, the combination of intensive agriculture, high population densities in small local areas, dry climates, cattle herding (the main cattle herds in Ghana are located in northern Ghana) and the lack of rural investments resulted in severe soil erosion and eventual desertification (Pearce 1992). Benneh and Agyepong (1990) describe desertification as the ultimate form of land degradation where the biological, productive capacity of the land or water is destroyed. While there are no permanent deserts in Ghana, there are several areas where desertification is a serious threat. The total area at risk from desertification in Ghana has been estimated at 83,487 square kilometres, or 35 percent of the total area of Ghana. These areas include the Coastal Savanna, the Upper West Region and the eastern part of the Northern Region (see Figure 7 for the map of the Regions of Ghana).

The main problem region is the Upper East Region. Here intensive soil erosion occurs from the combined effect of population concentrations of over 200 people per square kilometre on the watershed areas of rivers (due to the incidence of onchocerciasis or river blindness, a disease which can cause blindness in the river valleys), the dispersed settlements practising compound farming over large areas, the very large herds of cattle, the fragmentation of land under the traditional system, the use of fire for clearing land, and the cutting of wood for fuel. In many areas, denuded landscapes characterised by intensive sheet, rill and gulley erosion have replaced the natural vegetation of short grass and occasional shrubs and trees.

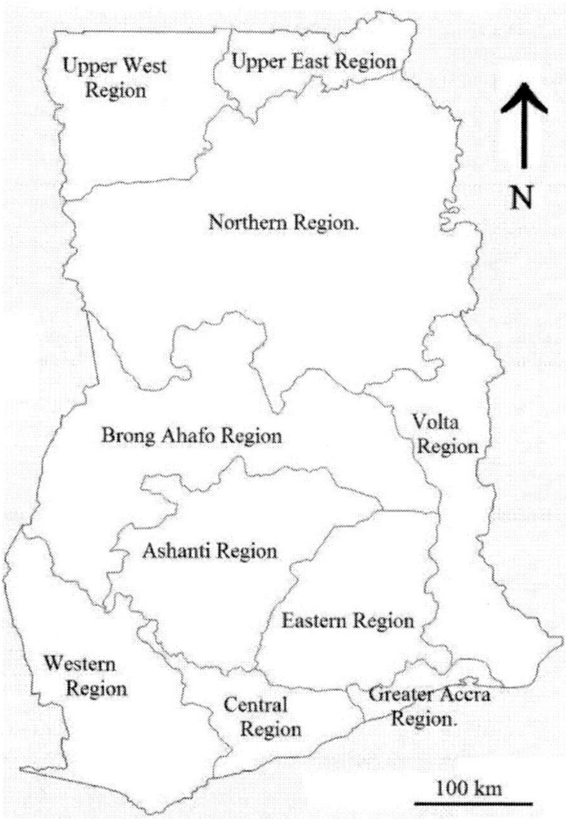

Figure 7. Administrative Regions of Ghana.

5.2. Forestry, Deforestation, Soil Erosion and Social Effects

Historically, forestry was a major factor behind deforestation, although it was less important economically than cocoa (Howard 1978). The primary forest areas of Ghana declined from 74,400 square kilometres in 1977 to 18,500 square kilometres in 1987 (Benneh and Agyepong 1990). Although some believe that over 70 percent of deforestation in Ghana was due to shifting cultivation (Agyemang and Brookman-Amissah 1985), the forestry industry was also a major influence (Norton 1987; Friends of the Earth Report 1992). The main exploited woods were mahogany *Khaya ivorensis* and *K. anthotheca*, obeche, wawa or santa *Triplochiton scleroxylonm* K.Schum, sapele *Entandrophragma cylindricum*, dahoma or ekhimi *Lovoa trichicloides* and guarea *Guarea cedrata* (Harrison Church 1980). Exports of timber increased from 84,950 cubic metres in 1888 (Dickson and Benneh 1988) to 236,000 cubic metres in 1951 (Dunn and Robertson 1973:.81).

The logging industry was done largely by government-granted concessions operated by large private companies, frequently on land previously utilised by small-scale farmers and/or fuel wood cutters. Selective logging was employed to extract the needed trees and the logs were floated down the rivers such as the Offin (a tributary of the Pra) and the Ankrobra, or transported on trucks. Loggers destroyed many different trees to extract the valuable species. Many areas were repeatedly logged, giving the vegetation no time to recover (sometimes the felling cycle was reduced from 40 years to 15 years).

There were numerous incentives for the continual destruction of the forest. These included the non-competitive pricing of wood, and inappropriate and ineffective tenurial arrangements. Even though the state was the legal owner of the forests, control over excess logging was rarely enforced, and the forests were thus open access resources. Private logging companies received concessions to exploit timber, but their subsequent activities were rarely monitored (Environmental Action Plan 1988). The demand for fuelwood intensified the tree felling. Historically, fuelwood comprised leftovers of logging, fallen branches, deadwood and land cleared for agriculture. Later, demand for fuel wood increased, outstripping the supply from traditional sources, and encouraging cutting of living trees. In 1990, the domestic consumption of fuelwood was estimated at 5,547,200 cubic metres, while the consumption of charcoal was estimated at 1,992,400 cubic metres (Benneh and Agyepong 1990).

The decline in forest stands meant that the collectors of fuelwood walked further for suitable supplies. Intensive logging and the reduction of regeneration periods degraded soil fertility (due to reduced litter layers, the exposure of the soil to the sun's rays, the erosive effect of raindrops, and the invasion of grasses that provided less effective protection than trees) and thus forced food farmers (who were already fighting a battle against competing interests such as cash cropping and increased demands for food) to move their farms deeper into the forest, increasing foraging and deforestation.

5.3. Urbanisation, Population Distribution and Soil Erosion

Urbanisation and population growth have been major factors for environmental change (Campbell 1998). During the colonial period, the population of the Gold Coast colony increased sharply after 1901 from 1,948,000 in 1901 to 3,164,000 in 1931 (Austin 2008), 6.7 million in 1960 and 24,233,431 in 2010 (CIA 2012). The urban population percentage increased from 5 percent to 7.7 percent in 1931 (Jedwab and Moradi 2011), and from 23.1 percent in 1960 to over 50 percent in 2010 (Owusu-Ansahm 1995; CIA 2012). Urban settlement generally implied denuding the landscape of vegetation. As noted by Benneh and Agyepong (1990) there is hardly a settlement in Ghana without inter-house erosion. This existed mainly in low-income areas, where the local actors have little access to repair facilities, and construction materials were mud, stone, low-grade cement and corrugated metal, on a base of exposed soil. The fringes of urban areas expanded to meet the needs of the influx of immigrants from the rural areas, denuding these areas of vegetation, and encouraging runoff and erosion on open yards, dirt roads and paths. Inter-house and street erosion were serious problems that contributed to social problems, such as the loss of homes, roads and multipurpose open spaces, and scenic and physical dereliction The heavy, convectional rainfall, coupled with the expansive effect of heat on the asphalt, concrete and dirt surfaces, readily degraded roads, streets and sidewalks. Poor drainage facilities also allowed flooding (both on flat topography such as in the city of Accra, and on hilly, rolling topography, such as that of Kumasi) and the blockage of gutters and drains, these reducing streets, open areas and markets to crowded places of filth, squalor and seasonal mud (E.A.P. 1988). Inefficient waste disposal methods exacerbated the unfavourable effects of erosion, flooding and deposition.

These problems were not limited to the large urban areas. Smaller urban and rural settlements also suffered from severe erosion problems, compounded by the cultural dualism which encouraged much greater investments in infrastructural development in the large urban areas, and neglected the smaller urban and rural settlements, the result of the lack of participation of poorer actors in the power structures (Benneh and Agyepong 1990). An example of a rural area facing increasingly disastrous erosion with inadequate development investments was the village of Agromeda, situated about 80 kilometres northeast of Accra. Populated since 1892, there was gulleying with depths of 61 to 152 centimetres. The land was bare and erodible due to clear weeding, sweeping, rainwater gushing from the roofs, and runoff from the tarred roads. This led to the exposure of the foundations of buildings, cracked and falling walls, and abandoned homes (Benneh and Agyepong 1990).

CONCLUSION

The literature cited above gives a brief, selective perspective on the perception of environmental change during the colonial period in the forested areas of Ghana. The traditional agricultural system, based mainly on shifting cultivation and/or bush fallowing, was clearly suitable to the needs of the time, but a larger population and a new international reality compelled a conversion to a dual agricultural system. The extent of environmental degradation depended on negotiation between the various stakeholders and the environment through time. While it may be possible to discern a general trend of increased environmental degradation with population increase, there was also a patchwork of diverse possibilities as local people adapted and changed for livelihood sustenance and environmental durability. The next chapter will examine these ideas in a case study from within Ghana.

Chapter 6

Soil Erosion in a Ghanaian Landscape

6.0. Introduction

This chapter examines soil erosion in a Ghanaian landscape, namely a region of the coastal savanna, focussing on the role of landuse, topography, soil desiccation and waterlogging in the general development of soil erosion. Based on the issues discussed in earlier chapters, the case study looks at the physical bases of soil change and social issues such as power among land users, social structures and actor volition, livelihoods, gender and social change. The appendices present the details of the field survey, including references to literature.

6.1. The Importance of Topography

The impact of topography and erosion on stakeholder competition for land and local rural decision-making requires more study in Ghana. Therefore, the consequences of such interactions are largely unknown. This omission is important for West Africa, a region appraised by contested narratives of deforestation and landuse impacts on vegetation and stakeholder conflict. Topographical features, especially slopes and elevation, present an important issue for landuse in rural Africa, due to the preference among variable stakeholders for gentle slopes (Scoones et al. 1996; Brouwer and Powell 1998; Russell 1999; International Food Policy Research Institute 2003; Powell et al. 2003; FAO 2005). Consequently, not only does landuse on steep slopes reflect the interplay of stakeholders' space access through time, it may also contribute to socio-environmental coevolution, biogeographical change and soil erosion (Tiffen et al. 1994; Iddrisu and Tropp 1995; Campbell 1998). Although assessments have been made of rural livelihoods and cultural biogeographies in areas of topographical variation (Taddesse 2001; Campbell 2004), and power interplay has been cited as a factor for landuse (Clay and Lewis 1990; Barbier 2000; Lemly et al. 2000; United Nations Convention to Combat Desertification 2002; Karar 2004), few studies examine the factors of slope and human behaviour in intensive contexts where both steep, hilly areas and gentler plains may be compared (Campbell and Palmer-Jones 1999; Ryszkowski and Jankowiak 2002). There are also few studies of the contribution of this dynamic to the complex features of an African rural landscape, which may include fragmented forest, shrubs, grasses and rivers, and people engaged in multiple activity livelihoods (Campbell 2005a).

In Ghana, the settlement of Accra (the largest and capital city) by the British colonial power was largely due to the flat, coastal topography (Figure 8). This influenced the movement of the capital of the Gold Coast colony from the city of Cape Coast (Dickson and Benneh 1995). In this area, the origin of the savana vegetation on the otherwise forested coast is blamed on rainfed farming and/or the naturally dry climate cycles (Carson 1985; Elliott and Campbell 2002; Bediako 2004). It now houses the largest population concentrations in the country, with substantial numbers of landusers from other parts of Ghana (Dickson and Benneh 1995; Campbell 2005b). It is also the main ecological footprint of Accra, utilised for farming, water extraction, quarrying and suburban settlement (Carson 1985; FAO 2003). Due to its dry climate and gentle topography, there is large-scale irrigation farming (Shirazu 1996; Opoku-Ankromah 2000; Ghana Ministry of Agriculture 2003).

This coastal savanna, the driest region of Ghana (Figure 6) is divided roughly into two vegetation and topographical types: flat, grassland east of Accra, and more hilly coastal scrub to the west of that city (Dickson and Benneh 1995). The coastal scrub is denser, with a more dynamic patchwork of tree stands, shrubs and fallow farmland. This area is also a more intensively utilised mosaic, with competing interests in building landuse, small and large scale farming, hunting, sacred grove preservation, firewood harvesting and rural settlements (Elliott and Campbell 2002; Campbell 2005b). Due to the strong deforestation narrative in the area, one of the strongest in Ghana (Choin 2002), there are important unanswered questions concerning the biogeographical impact of landuse activities, the influence of topography on these issues and the extent of soil erosion.

6.2. The Case Study of the Coastal Savanna

The study area was selected, as part of a larger study, as typical of the coastal scrub, with rolling hilly topography, covered by a mixed tree, shrub and grass vegetation mosaic (Figure 8). The objectives of the field study were to document and analyze (1) the role of topography in landuse; (2) the preference and access abilities of the various stakeholder groups to topography; and (3) the consequences of these dynamics for landscape vegetation patterns. There were two hypotheses: (1) topography influenced the landuse practices in the area; and (2) these relations were based on power play between stakeholders.

Situated about 20 kilometres southwest of Accra, there were strong links with markets and business centres of western Accra. There were six settlements in this area, in order of decreasing size: Bortianor, Oshiyie, Kokrobite, Tuba, Aplaku, and Langma. The total population was approximately 8000 in 2000 (Census Reports 2000). The study area was under strong landuse competition, as local actors, mostly farmers, hunters, woodcutters and traders, competed with each other and land developers for economic and residential spaces. The dry climate introduced an additional dynamic, as farming favoured lowland and moister areas, and was the factor for the location of a large irrigation project in the area. Social and ecological field methods, and GIS based quantitative analysis documented the socio-environmental relations in the area. Soil type was verified according to sampling and the Soil Research Institute Classification (1995). The correlation of slope and vegetation density (percentage of trees, shrubs and grasses) in upland areas (above 80 metres) and lowland areas was calculated using linear regression at the 0.05 level of significance.

A SPOT satellite image (1996) provided total coverage of the area. Aerial photographs dated 1960 and 1986 provided partial coverage. The vegetation recorded in the field survey and identified in the aerial photographs was classified into tree/shrub and grass/farm. Feature maps were created using ESRI GIS for both the satellite images and the aerial photographs, to discern landcover change (see Appendix 1 for the results of the landcover change study). The average slopes of the features in the field survey and the feature maps (vegetation, farms, building plots) were calculated using ESRI GIS and IDRISI for both sets of feature maps (1960 and 1996) using digitised contours.

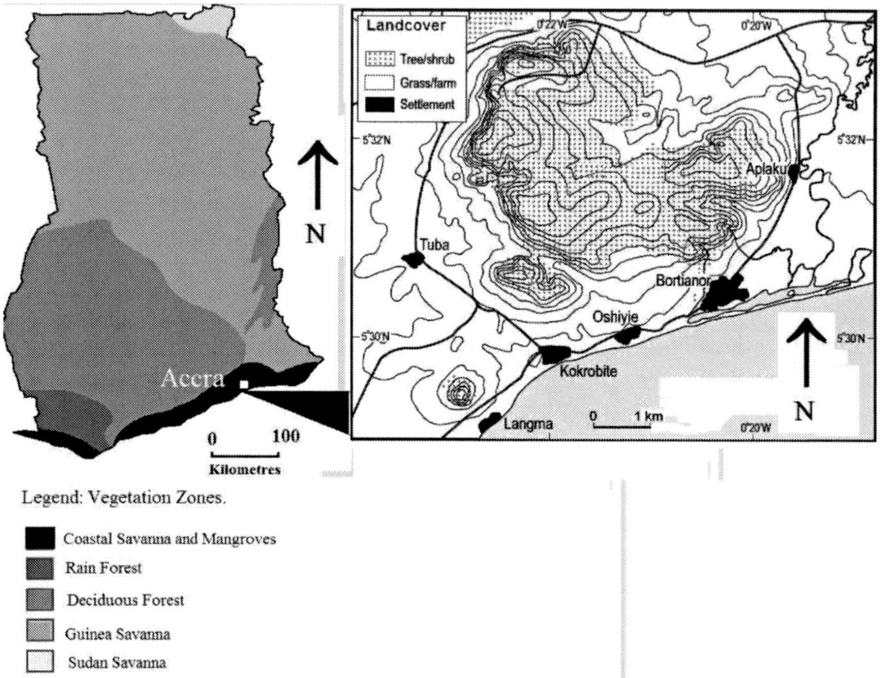

Figure 8. The Study Area of Coastal Ghana.

6.3. FARMING LANDUSE 1940 -1960: THE CULTIVATION OF LOWLAND PLOTS

During the period 1940 to 1960, nearly all farms were located on the flat lowlands near the villages. There were a few farms on the steep, upland slopes, but these slowly declined. The decline in upland farms (especially tomato cultivation) during this period was blamed on declining rainfall, as most farms were rainfed, cultivated according to the seasonal sequence listed in Figure 9. These statements were partly supported by the rainfall data obtained from the Meteorological Services Department. Although this data showed only gradual decline, there was extreme variation in annual rainfall, with large cycles of three to four years.

Close examination of the aerial photographs showed barely visible margins of farms in some upland areas, all in areas with moderate slopes ($< 6^{o}$). The local interviewees supported the existence of these old farms. Older people, especially men, argued that farms in the pre 1950-60 period were scattered in upland areas. Farm location was based on the bush fallowing land tenure system. Plots were cultivated for two to four years, and then left fallow for up to five years for natural soil fertility recovery.

By 1960, the majority (80 - 85 percent) of cultivated plots were still located on flat, lowland areas, less than two kilometres from the villages. Areas over 80 metres in elevation over the flat plains were largely uncultivated (93 percent), and covered by dense tree/shrub vegetation with continuous canopy. No areas of slope more than 15° were cultivated, with 95 percent of farms being located on slopes of less than 8°. The lack of landuse intensity was evident, as more than half of those identifiable from aerial photographs were located more than 100 metres apart. Men were more tolerant of steep slopes and gentle uplands where access paths had steep inclines. Females cited problems, especially energy expenditure, field childcare, pregnancy and lower access to the best areas due to the superior power and status of male farmers in land assistant labour acquisition.

Month	Climatic Season	Farmer Decision Making
January	Peak of dry season.	Too dry for farming.
February	Start of transition to rainy season.	Too dry for farming.
March	Moderate rainfall.	Variable rainfall. Clearing starts.
April	Rainy season begins.	Clearing, ploughing and early planting.
May	Further increase in rainfall.	Planting must begin for good crop.
June	Peak of rainy season.	Main period for crop growth.
July	Beginning of minor dry season.	Beginning of harvesting.
August	Peak of minor dry season.	Most harvesting ends; new ploughing.
September	Beginning of second rainy season.	Planting starts.
October	Peak of second rainy season.	Main period for crop growth.
November	Transition to dry season.	Harvesting.
December	Dry season.	Too dry for farming.

Figure 9. Farming and Climatic Seasonality.

6.4. Farming Landuse 1960 -1996: Competition for Lowland, Marginalisation to Upland Slopes

During the period 1960 to 1996, there was increased cultivation of steeper slopes by both men and women. The farmers adapted to the dryer upland slopes by cultivating drought resistant crops (maize *Zea mays*, and less frequently cassava *Manihot utilissima* and sorghum *Sorghum guineense*) and using the lowlands for more drought sensitive crops, such as tomatoes *Lycopersicon esculentum*, peppers *Capsicum annuum* L. and groundnuts *Arachis hypogaea*.

The first reason for the increased use of upland, steeper slopes was the population increase from 1972 (1948) to over 8000 in 2000 (Government of Ghana 2000), due to both increased birthrates and immigration from other regions of Ghana. This increased the number of farmers, despite slight declines in farming popularity. Therefore, due to the general shortage of available farmland, during 1980-1982, the farmers' first coping strategy was to replace bush fallowing with relatively permanent plots. For security in their other activities, their second coping strategy was to cultivate scattered rainfed plots, as scattered farms (e.g. two or three plots of the same size rather than one larger farm of their combined area) offered more security against financial trouble and debt. As rainfall declined, the farmers tried to avoid the steeper, drier slopes, but mostly failed due to the population pressure on the preferred flatland. The increase in elevation and slope of local features is shown in Figure 10.

Landuse	**1960**		**1996**	
	Elevation (m)	Slope (.°)	Elevation (m)	Slope (.°)
Village	11	5.1	11	5.0
Tree/shrub	48	15.1	69	18.9
Grass/farm	21	5.8	38	8.9
Building plots	11	2.7	54	5.1

Figure 10. Average Slope and Elevation of Landscape Features.

A third coping strategy, mostly where multiple-plot farming was practised on unsuitable slopes, was networking in other livelihood activities. Other alternate activities for male farmers were woodcutting and trading (24 percent), hunting of small antelopes (six percent) and fishing (22 percent). Urban jobs, wood trading and transport assistant ships were also undertaken

(27 percent). Female farmers would resort to part time fish smoking (29 percent), trading (31 percent) and urban jobs (17 percent). However, these decisions were also tempered by the general belief that natural resources (wildlife, fish, soils and plants) were under slight or serious decline due to intensive use.

A second reason for the increased cultivation of upland, steeper slopes was the loss of large areas of lowland farmland to irrigation farming. In 1983, an irrigation project was set up, under the Weija Irrigation Company, which was owned by the Government of Ghana and supported by European Union funds. The project covered 200 hectares of lowland around Tuba and between that village and Kokrobite. The Weija Development Company obtained land from the traditional authorities for irrigated agricultural development in the early 1980s, without the consultation of the local farmers and forcibly evicting many farmers from some of the best land in the study area. Serious conflicts resulted including actual violent protests from the affected farmers, several of whom lost crops immediately prior to harvesting. Local farmer involvement in the project hovered around 100 for the period 1983 to 1996, with only a few women joining due to the heavy work involved (pipe carrying, and year round digging), the presence of dangerous reptiles attracted by moisture, childcare and household work, cultural norms favouring men in access to money lending, hired farm help and tractor ploughing. Although introduced to enable dry season cultivation (three or four seasons instead of the two allowed under the double rainfall maxima) and the increased cultivation of drought intolerant crops such as tomatoes, the high cost of electricity for irrigation pumping forced farmers either to leave the project or increase their participation in other activities (rainfed farming, hunting, trading, fishing and/or urban work).

A third reason for the increased cultivation of steeper, upland slopes was building landuse by developers, a powerful stakeholder group, on the best flatland especially after 1990. Building developers preferred flat land, although some plots were located on steeper slopes (15°+). The areas cleared and marked for building had average slopes of less than 6° and covered about 25 percent of the preferred farmland in the area. Large areas of gently undulating farmland were acquired from the traditional authorities, usually the chiefs of the villages and not the farmers, as the latter would likely be reluctant to give up land. With lowered agricultural outputs, land sales were deemed more beneficial than granting land to farmers. Plots of 1500 square metres could be sold for $1000, especially those on good land (flat, well drained, near roads). In many cases, traditional rulers removed farmers who had occupied land owned for several generations. Significant landuse conflicts occurred during 1993 to

1996, in the Bortianor and Aplaku areas, as several farmers refused to vacate farmed plots.

Clearance involved not only the removal of the vegetation and topsoil of the plot, but also that of the surrounding area (sometimes to a distance of 15 to 20 metres), and wide swaths of vegetation to the nearest road. The total cleared area could destroy up to twenty farms, and destroy their agricultural value, even if the temporary road connections were later abandoned. Significant periods elapsed (1 – 7 years) between the purchase of land (which resulted in the insertion of concrete block markers on the plot corners) and land clearance by the heavy tractors or graders, during which period farming on the plots ceased. Hence, dense fallow vegetation generally occupied such areas.

6.5. Vegetation Change in Upland and Lowland Areas

There was substantial vegetation change between 1960 and 1996 (Appendix 1). The average slope and altitude of dense vegetation increased markedly. The main changes were a decline in dense tree/shrub areas, especially in lowland areas and the increased fragmentation of such vegetation, in favour of grassed patches. Tree losses, due to intensified cultivation and fuel wood extraction, were highest in the upland areas of the hinterland and Aplaku, these areas being under largely contiguous forest in 1960. Lowland areas, especially Tuba, Kokrobite and Oshiyie, showed slight change and even some increased tree stands in the margins of villages and graveyards.

There were tree stands located on the flat ground surrounding each village, maintained for shade, occasional wood harvesting through branch and root coppicing, and fruit picking (mangoes). These stands showed only a slight decline (101,233 square metres (1960) to 95,367 square metres (1996). Unprotected tree stands in lowland areas showed a sharp decline (40 percent). In some areas, there was almost total loss of contiguous forest stands, especially in the lowland coastal areas surrounding the tree stands on village margins. The actual area of the villages increased only slightly (5 – 7 percent), replacing some of the surrounding tree stands with buildings. Dense tree/shrub vegetation in the margins of farms (usually 1 to 5 metres around farms) declined in lowland areas (8 – 10 percent), as some farms expanded in area.

Those in irrigated areas showed a slight increase in area (8 – 9 percent). In upland rainfed farms, the marginal vegetation declined (10 – 15 percent).

There were four sacred groves in the area, located on flat land near the villages of Oshiyie, Kokrobite and Bortianor. Sacred groves were areas set aside in the 19th century, where religious idols were located, with substantial areas protected from all landuse by traditional religious leaders. The administrative and spiritual power of the local chiefs supported these edicts. Despite scattered violations by farmers and woodcutters, especially along the margins, there was little areal change between 1960 and 1996. Total sacred grove areal change was from 193,166 square metres (1960) to 165,531 square metres in 1996, a loss of 15.3 percent. Interviewees blamed the pressure of farmland decline for the losses. The vegetation was similar to the inland deciduous forest, also with local tree species such as the neem *Azadirachta indica* and mango *Mangifera indica*.

6.6. Soil Erosion, Desiccation and Waterlogging

There were four soil orders (United States Department of Agriculture USDA Classification) in the area: Alfisols (Haplustalfs and Natraqualfs), Entisols (Ustipsamments, Ustorthents and Endoaquents), Ultisols (Haplustults and Paleaquults) and Inceptisols (Endoaquepts) (Appendix 2). These were mixed into complexes with local names. There was a high association between soil type and slope/elevation. With some variation, upper slope soils were generally paler and reddish, well drained and hence much more prone to desiccation. All farmers preferred low elevation soils, which were largely dark with clayey and loamy horizons. However, many building developments were located on the latter soils, due to their flat, lowland topography, an important factor for conflicts as noted by all respondents.

The fertility difference between the red and black soils became especially relevant to farmers in the period 1980 to 1985. Then, due to population and landspace pressure, permanent plots replaced shifting cultivation/bush fallowing, starting competition for good soils and flat land with farmers, developers and the irrigation project. A particular handicap of red soils was their sandy horizons. When tractor clearance became the norm, as larger areas were needed to be cleared to increase production, the ploughs churned the sandy and infertile subsoil into the topsoil, decreasing soil fertility and increasing the farmers' awareness of soil profiles, fertility variation and the need for fertilisers.

Erosional features in the study area were of two main types: accelerated erosion on farmlands and sheet, rill and gulley erosion on fallow areas. Accelerated erosion on farmland was a serious problem on the ploughed fields of the study area. The amelioration of this phenomenon cost the farmers a great deal of time, energy and money. One factor was the clearance of soil protecting vegetation, necessary to allow tractors to plough plots. Plants such as Kanto *Zanthoxylum xanthoxyloides*, Haartso *Fagara zanthoxyloides* and Neem *Azadirachta indica* were very difficult to clear, so many farmers, mostly women hired special cutting teams to clear the bush and dig up roots. The cost of this work in money, time and energy, influenced some farmers to avoid sites where such species were dominant. However, the thick and difficult vegetation offered the best protection against soil erosion during fallow periods. Light, easily clearable vegetation offered little protection against early rain erosion, before the recovery of deciduous vegetation. The farmers regarded this dilemma as ironical: "the areas that are easy to clear, also bring more soil problems."

Most plots were ploughed by tractors in the direction of the slope. This created erosion problems. Accelerated erosion of this type in ploughed fields is a well-noted problem in the literature. Nir (1983: 87) notes:

> The effects of ploughing depend heavily on the angle between the direction of the furrow and that of the slope. Ploughing from the top of a slope to its foot may technologically represent the simplest manner, and it certainly is used. This is however the most dangerous manner of soil cultivation: the furrows cause run off. This type of intervention is perhaps the single most important agent of erosion induced by man.

Therefore, erosion moves from ridges into furrows and down the slope. Ridges and furrows were made during the rainy season, to avoid waterlogging. However, heavy rain destroyed ridges and hill crops and caused waterlogging, especially in low-lying sections of the fields. Where the ploughing did not follow the slope, but cut across it, or where the land was flat, erosion still occurred. Ridges where the fertiliser was deposited were eroded with deposition occurring in the furrows. This necessitated continuous work repairing the ridges, requiring on average twelve visits to the farm per month. Other visits were for weeding (9 visits), fertiliser application (3 visits), bird scaring (2 visits) and inspecting (2). In some cases, the level of erosion was severe, so more ploughing was needed. When such a second or third ploughing was done after planting, the whole of the first crop could be lost. In such cases, the farmers would undergo an extremely traumatic experience, borrowing money from moneylenders or relatives to pay the tractor owners for further ploughing, to buy seeds, fertilisers and perhaps pesticides, and hoe the field all over again. These activities were done with extreme urgency, because of the limited rainy season

Soil erosion under fallow vegetation was also important. In the survey conducted on vegetation analysis, every quadrat was examined for soil erosion occurrence (see Appendix 3). The following facts were noted: (1) soil erosion was greater under grasses than shrubs; (2) long fallow vegetation was a much better protection for the soil than short fallow vegetation; and (3) soil erosion was very serious on steep slopes, in fact, in some cases, farming was impracticable. Sheet erosion was measured by root exposure, before and after rains. Deposition would also occur at the base of grass roots, giving rise to heaps of soil, gravel, stones and small boulders. In some long fallow areas, especially on very steep slopes, gulleying was common even though the soil was under dense cover.

Soil stoniness was a measure of soil erosion, (note, in order of size boulders 20 cm+; stones 2 - 20 cm; gravel 2 -20mm; Ahn 1993: 57). All the upland soils were extremely stony. The thin soils, generally classified as entisols, were sometimes fertile, especially when black and loamy. However, the stones caused serious problems for farmers, increasing the need for hoeing and exacerbating the likelihood of desiccation. Frequent comments were: "One day the rain will come so heavily that our crops will be washed away, after the drought has killed them"; "The soils here are not strong"; "You can't even climb the hill without falling down"; "You have to walk carefully, if you fall holding a cutlass you can cut yourself"; "You have to get off your bicycle much earlier when ascending a hill, because of the deep gullies."

Upland red soils were increasingly prone to desiccation, due to several factors: (1) the declining and extremely variable rainfall; (2) the relative humidity, which gradually declined during the dry season; (3) the relief which caused significantly greater ground water loss from upland areas; (4) the soil types of the upland areas, which were stony and shallow, with little moisture retention; (5) the cool strong sea breeze which dried the hill slopes; (6) the severe sheet erosion on cleared upper slopes, which exposed subsoils to direct sunlight; and (7) the decline in vegetation cover, due to rainfall decline and variability, vegetation clearance in upland areas, and intensified cultivation (Appendix 4). Human activities exacerbated the effects of these natural factors. These included: (1) the burning of large areas of shrub and grassland on hillsides and hilltops; and (2) the absence of any form of soil water management, such as terracing, that would attempt to conserve moisture.

One solution to desiccation was the head loading of water up the slopes, a practice that occurred for 40 to 50 years. The water was carried principally for tomato crops. Men usually used 30 gallon metal barrels, while women used smaller 30 litre containers. The men would on average, make four trips up the hills per day (two in the morning and two in the evening). Women tended to "go up" more often with their lighter load; six or seven trips daily were the norm. The hard manual labour involved in the clearance of thick bush (thicker than the present shrubs) and carrying water deterred most women from farming the hillsides.

Desiccation forced farmers to abandon several farms located on the higher slopes. Crop losses included the drought tolerant maize and most of the perennial cassava. The year 1995 witnessed massive losses of tomato crops on the upper slopes, (due to gaps between rainfall), and even on the flatlands (due to extremely intensive storms which caused waterlogging). Some lowland areas were permanently flooded, even during the dry season, due to the combined effects of excessive water drainage from the irrigated areas, high rainfall during the rainy season and the thick clayey soils. Nine rainfed farms, located on flat land near the irrigated areas were abandoned, due to waterlogging and crop losses.

To verify the soil desiccation problems of the farmers in the study area, two sets of tomato and maize farms were compared: three on the hill slopes and three on the plains. Figure 11 displays this information. The two crops were selected because of their differing moisture requirements. Tomatoes grow best in the moist deciduous forest with annual rainfall of 1250 to 1750mm, while maize can tolerate 200 to 900mm annual rainfall. Thus, maize is more drought-tolerant and can tolerate deep soil desiccation. However, tomato is less resistant to waterlogging. The growing period for maize is usually about 100 days. The growing period for tomatoes is about 80 days, following 4 weeks in the nursery.

Over a three-month period (May 20 to August 20) crop growth of the crops was monitored. On the upland farm, the farmers carried buckets of water up the slope for the tomatoes during the nursery stage. After 25 days, the tomatoes were transplanted. At the end of the observation period, the entire tomato crop on the hill slope was decimated by desiccation, despite the efforts of the farmers. There were sporadic attempts to carry water by bucket from the village. The maize survived, with the loss of a few plants. The farms in the valley experienced only minimal losses due to desiccation and occasional waterlogging.

The incidence of soil desiccation and drought robbed the peasant farmers of funding from the Agricultural Development Bank, another power actor in the area. In 1989, at the beginning of the farming season, the ADB gave loans to 15 tomato and pepper farmers from Kokrobite, with a planned expansion if the selected farmers were able to pay back the loans. However, due to a severe local drought, the crops failed and a dispute evolved between the farmers and the ADB over the farmers' inability to produce enough crops to pay back the loans. Finally, the farmers requested the Bank officers to inspect the farms, to verify their argument that drought was the factor for their ruin. Although after inspection the ADB officials agreed with the farmers, they concluded that further loans would be too risky. This forced the farmers to return to local sources of loans, especially fish smokers and traders.

Farm	Location	Slope	Crop	Hectares	Losses(%)	Reason for Loss
1	Hillside	22^0	Tomatoes	0.303	95	Desiccation and Erosion
2	Hillside	23^0	Maize	0.324	18	Desiccation and Erosion
3	Hillside	22^0	Maize	0.202	10	Desiccation and Erosion
4	Plain	1^0	Tomatoes	0.486	22	Desiccation/ Waterlogging
5	Plain	3^0	Maize	0.445	2	Desiccation and Pests
6	Plain	1^0	Maize	0.485	2	Desiccation and Pests

Figure 11. Crop Location and Soil Desiccation.

Waterlogging was confined to the flat areas of the study area, especially depressions, soils with clayey and/or indurated horizons and those hardened by long droughts (see Appendix 5 for more information on waterlogging in the area). The most serious years for waterlogging were 1986, 1990, 1993 and 1995, when between 10 and 50 percent of the crops were lost. No tractor could enter the waterlogged fields without sinking into the mud, and no method was devised for draining the waterlogged farms. In some areas, whole farms were abandoned due to waterlogging. The chronology of one farmer's problem with soil waterlogging (Adawso soil series) is shown in Figure 12.

Time	Waterlogging and Tillage Process
Day 1	Tillage by tractor
Day 2	Creation of ridges and furrows by hoe begins
Day 3	Heavy rain. Work is suspended.
Day 4	Walking on field creates depressions 25 cmm deep.
Day 5	Walking depressions 8 cm. Hoeing starts.
Day 6	Rain. Depressions 14 cm. Hoeing stops.
Day 7	Depressions 8 cm.
Day 8	Depressions 5 cm. Hoeing continues.
Day 9	Hoeing continues.
Day 10	Light rain. Depressions 6 - 7 cm. Hoeing continues.
Day 11	Planting begins.

Figure 12. Farm Waterlogging and Ploughing Problems.

The attraction of potentially dangerous animals, such as snakes and big lizards was a problem. Sightings of such animals increased by 400 to 500 percent during waterlogged periods. Mice were attracted, not only by water during the dry season, but by crops. The colonies of mice also spread to the rainfed farms during the rainy season, feeding on half grown tomatoes and garden eggs. Dangerous chemicals were used to kill the mice.

6.6. Soil Fertility Losses

One of the main effects of soil erosion, desiccation and waterlogging was soil fertility loss, especially after the transition from shifting cultivation to permanent plots and irrigated agriculture in the 1980s, which brought fertility decline over three to five years. The annual fallow period of 5 months (the dry season) was sufficient for some soils to retain fertility. The farmers were able to estimate the need for fertiliser or fallow periods, and impacts of intensification of cultivation (see Appendix 6 for more information on farmer soil knowledge).

Most of the farmers only began to use fertiliser on a regular and intensive basis in 1989 to 1990, annually using as few as 1 or 2 bags (50 kilograms each) for fertile, loamy black soils, or 10 to 15 bags for the sandy, red soils. There was a 70 to 80 percent increase in the quantity of fertiliser purchased, and a 700 percent increase in the proportion of farm expenditure spent on fertiliser, financed largely by alternate activities. Appendix 7 shows some of these activities, and the problems associated with them. In 1991, many farmers added cow and green manure as fertliser (mostly old, harvested maize or rattle weed *Crotalaria retusa*).

6.7. Coping Strategies, Gender Structures and Actor Volition

The coping strategies employed by farmers for additional income were enabled and constrained by local social structures managed by individual actor volition, wider national level political and economic structures, and power relations, such as language, education and urban development. Gender specificity for certain livelihood activities was a powerful structure, with enforcement based on peer pressure, social embarrassment and "teasing". The strength of this structure was nevertheless sufficient to keep people in line, despite no real punishment for "deviants."

The hunting/meat and fishing industries were strongly gender specific, with a male half (killing animals and sale of skins) and a female half (processing and sale of meat). Older interviewees argued that in the 'early days' of the past centuries, much larger animals than duikers were hunted (elephants, river hogs, leopards). Hence, hunting developed as male profession. Also discouraging for women was their disproportionate share of housework. All the people who fished at sea were males, and the fish sellers and smokers were all female (and hence the main moneylenders). The factor behind the male dominance of fishing was a perception of strength: the hard work of rowing, managing motors, dragging nets and boat launching required minimum numbers, as profits were shared. In addition, fishing was an 'all day and some night' activity; this interfered with female household duties. Fish smoking and selling were female dominated, due to the traditional perception of food preparation as a female activity, and the possibility of locating the business areas close to the home.

Men dominated cattle herding, but cattle ownership was gender neutral. Farming and fuel wood extraction were gender neutral. There were some gender-based differences, as men farmed larger areas (mostly double the area of the female farms) and cut heavier roots and branches.

Actor initiated change was important, as the strongly defined gender specificity described above was undergoing change. The change was largely due to changes in the structures that dominated these relations. Male dominated activities had declined, due to environmental degradation and the development of alternate activities. The decline in wild mammal species affected the profitability and status of hunting. The reduced fish catches also diminished the status of fishing. Contemporary with these events was the rise in the status of moneylenders (mostly female) who assumed dominance over those actors weakened by environmental degradation and hence in need of loans.

The structural influence of children's livelihoods also changed. Parental influence and training declined as a factor behind the structures of gender relations. Training for boys and girls in traditional livelihoods either had declined, or was complementary to modern schooling. To many adults, "modern" children respected "book" knowledge and "foreign" cultural values more than the heritage of their parents.

Educated jobs superseded traditional apprenticeships. This may also eventually destroy the gender specificity of traditional livelihoods. Older people reminisced of their youth when superstitions surrounding sickness and death superseded "foreign" medical ideas. Peer pressure, fear and conscience would ensure the adherence of residents to the traditional values. The status of age ensured the promotion of ideas through the generations.

6.8. Summary: A Political Ecology Assessment

From the political ecology perspective, the non-farming livelihood activities encouraged complicated political and economic relations, creating new structures under continual reform and redirection, responding to complications in the environmental scene.

Power, economic variables and socio-cultural structures, all basic tenets of the political ecology paradigm, were strongly in evidence. Arguably, the coping strategies practised by local actors were evidence of the need for actor-oriented analyses of livelihoods in a political ecology framework. There were also structural influences, especially gender relations, competing landuse stakeholders such as the ADB, irrigation project and developers, and powerful local actors, such as the moneylenders.

The environmental variables, namely slope, elevation, soil texture, fertility, erodibility, desiccation and waterlogging, were analysable as structures, as they blended with social structures, as enabling and constraining factors for coping strategies and livelihood activity organisation. Coastal proximity created structural differences in the livelihood activities in certain areas (see Figure 8). The coastal villages relied more on sea-based activities like fishing, while those villages inland relied more on hunting for meat protein.

Therefore, a holistic perspective could see the processes of change as a duality; one way we see structures forming in socio-environmental interfaces and enabling and constraining individual action; looking another way we see individual actions enabling, maintaining and modifying these structures.

CONCLUSION

This book has examined a broad and integrated approach to the study of environmental change in West Africa and the case study from Ghana. The strong point of political ecology is the integration of relationships in politics and economics with the structures of ecology. These issues are vital, both for the strength of the perspective of political ecology and the possibility of solving environmental problems. The analysis of these issues from the perspective of political ecology will not necessarily offer an immediate solution to the externalities of environmental change. Rather, the improved understanding offered may strengthen a more enlightened approach to development problems that will at least serve to avoid the more serious problems that result from less sensitive analyses. Also, the implementation of conservation or reforestation schemes must take account of the wider social values related not only to environmental perception, but also to how people use the environment to survive, largely within the locally developed methods for conservation and management.

However, problems emerge for such a broad perspective, based on issues of coherence, relevance, compatibility and predictability. In this book, the weak points of political ecology include those emerging from the structuralist methodology of political economy, and the lack of ecological detail in the social science background. Hence, structuration theory and actor-oriented approaches complemented the structuiralist focus of political ecology. As was revealed in the case studies, structures are very important in an African context, as livelihood, gender and economic structures are vital means for local management of environmental resources. The concept of the coping strategy, based on actors and social and environmental structures, is analogous with structuration theory and gives a contribution that supports political ecology.

For the political ecology of soil erosion, there are problems not only with the "political", but also with the "ecology." Soil erosion is a huge topic, related to vegetation change, soil type, parent material type, topography and climate. Each of these variables substantially changes the ecology to the extent that the political ecology relationships change. In the case study, all these variables were manifest, and might be termed structures within a structuration theory framework. The question emerges, how does political ecology, itself dependent on changes in power relations between social groups, changes in structures used by local actors as coping strategies, changes in the power of social, economic and political structures versus changes in individual actors' powers of individualism and creativity, cope with changes in the ecology of the natural resource base? One solution is to increase the flexibility of the theoretical framework, giving full attention to the variability of each variable, whether social or environmental. In such an analysis, where both social and environmental variables (relations and entities) are the building blocks of livelihoods, the structuration theory is useful. The environment is also a structure that enables and constrains actors, is modifiable by them, and may modify their individualist and structured lives.

How to operationalise the results of these theoretical debates and apply them to the needs of local people depends largely on an understanding of what local people want from their environment. Therefore, wherever environmentally based activities are practised, cultural values and social needs, and ecological knowledge are important issues, identifiable as structures. Local people must guide environmental activities, for their stability to be possible over the short and long term.

Through the practice of shifting cultivation, the local people have demonstrated that they understand the need for socio-economic sacrifices to ensure environmental renewal and future benefits (a core argument of political ecology and some key definitions of sustainable development). Therefore, the question is not whether they understand the need for conservation, but whether after weighing the odds and appraising their present and future needs, they perceive such planning to be beneficial. This is a key emphasis in any development plan contemplating environmental conservation and renewal. Human values are environmentally relevant phenomena, because the manner in which people interact with their environment depends on their belief systems, what they perceive as important, their attitudes to environmental exploitation and their socio-culturally derived group identity. Value systems interacted with other aspects of the social context. Interpersonal communication, gender perceptions, marital relations, attitudes to age, child rearing, religion, status and traditional authority were values with powerful influences.

Individual decision-making, technology adoption and livelihood organisation, affected the acquisition of environmental knowledge by influencing the type of environmental contacts necessary for sustainable livelihoods. However, such structures were not static, but dynamic, gradually changing as the result of myriad individual decisions. Status change, livelihood insecurity, group communication, cultural resistance and gender relationship dynamics were issues that local people used to measure the success and appropriateness of projects and technology. In the study area, research suggests that these local appraisals were more appropriate and thus more accurate than the purely economic appraisals. The multiplicity of viewpoints on change is thus a relevant point for development planning. Hence, the utility of the political ecology approach depends on the documentation of societal power relations, the economic, political and socio-cultural structures from which these power relations emerge, the factors behind these structures, the extent of their rigidity or dynamism under variable actor oriented changes, and their relevance and relation to environmental character and change.

References

Abrahams, E. 1971. The hope of our cattle rearing industry. *The Ghana Farmer* XV(1):1-6.

Adams, W.M. 1990. *Green Development, Environment and Sustainability in the Third World.* Routledge, London.

Adejuwon, J.O., and Adesina, F.A. 1992. The nature and dynamics of the forest-savanna boundary in south western Nigeria. In Furley, P.A., Proctor, J. and Ratter, J.A. 1992 (Eds.) *Nature and Dynamics of Forest-Savanna Boundaries,* 331-352. Chapman and Hall, London.

Adimorah, E.N.O. 1995. Education and training of agricultural information personnel and agricultural information users in Africa. In Agricultural Information in Africa, 154-168. *Third World Information Services,* Ibadan.

Agarawal, B.1980. The woodfuel problem and the diffusion of rural innovations. Science Policy Research Unit, University of Sussex, 150. Cited in Blaikie, P. *The Political Economy of Soil Erosion in Developing Countries, 28.* Longman Group Ltd, Harlow.

Agarwal, B. 1991. *Engendering the Environment Debate: Lessons from the Indian Subcontinent CASID Distinguished Speaker Series,* No. 8, Centre for Advanced Study of International Development, Michigan State University, Michigan

Agarwal, B. 1994. Gender, resistence and land: interlinked struggles over resources and meanings in South Asia. *Journal of Peasant Studies* 22 (1): 81-125.

Agrawal, B. 1998: The gender and environment debate. In Keil, R., Bell, D.V.J., Penz P. and Fawcett, L. (Eds.), *Political Ecology,* 193 - 220. Routledge, London.

Agbesinyale, P.K. 1992. *Agricultural Expansion and the State of Ghana's Environment: the Need for Sustainable Development University of Cape Coast*, Cape Coast.

Agricultural Development Bank 1973. *Financial Report ADB,* Accra

Aguilar, A.C. 1995. *Priorities and Concerns of a Zarman Village in Niger, West Africa: a Preliminary Step in the Participatory Development Process.* University Microfilms International, Ann Arbor.

Agyei-Mensah, S. 1996) New perspective on fertility Situation in Sub-Saharan Africa. *Norsk-Geografisk-Tidsskrift* 50 (2):101-112.

Agyemang, V.K. and Brookman - Amissah, J. 1987. Agroforestry as a Sustainable Practice in Ghana. In National Conference on Resource Conservation for Ghana's Sustainable Development, Environmental Protection Council: *Conference Papers,* 2, 131 - 140.

Ahn, P. 1993. *Tropical Soils and Fertiliser Use.* Longman, Harlow.

Aidoo, J.B. 1996. *Tenancy and the Reform Debate in Ghana Royal Institution of Chartered Surveyors (RICS),* London.

Aikman, S. and Unterhalter, E. Eds. 2005. *Beyond Access: Transforming Policy and Practice for Gender Equality in Education.* Oxfam, Oxford.

Akiyama, T. and Larson, D.F. 1993. *The adding Up Problem: Strategies for Primary Commodity Exports in Sub-Saharan Africa.* World Bank, Washington DC.

Akuffo, S.B.1989. *Pollution Control in a Developing Economy: A Study of the Situation in Ghana.* Ghana Universities Press, Accra,.

Akuoku-Frimpong, H.1986. *The Need for an Effective Decentralised Framework to Promote Development from Below.* In Brown, C.K. Brown (Ed.) Rural Development in Ghana, 295 - 304. Ghana Universities Press, Accra.

Albee 1996. Beyond 'banking for the poor': credit mechanisms and women's empowerment. *Gender and Development* 4 (3):48-53.

Allaby, M. 1994. *The Concise Oxford Dictionary of Ecology Oxford University Press,* Oxford.

Allan, J.A. 1991. Global biogeography: who is responsible (Guest Editoral). *Journal of Biogeography* 18, 121-122.

Allen, R.1980. *How to Save the World: A Strategy for World Conservation.* Kogan Page, London.

Allen, T. and Thomas A. 1992. *Poverty and Development in the 1990s.* Oxford University Press, Oxford.

Allison, F.E. 1973. *Soil Organic Matter and Its Roles in Crop in Production.* Elsevier Scientific Publishing Company, Amsterdam.

Alo, C.J. and Pontius, R.G. 2004. *Detecting the influence of protection on landscape transformation in southwestern Ghana.* Clark University, Worcester.

Alsop, R.G., Khandelwal, R., Gilbert, E.H. and Farrington, J. 1996. The human capital dimension of collaboration among government, NGOs, and farm families: comparative advantage, complications, and observations from an Indian case. *Agriculture and Human Values* 13 (2): 3-12.

Altieri, M.A. and Anderson, K.A. 1986. An ecological basis for the development of an alternative agricultural system for small farmers in the third world. *American Journal of Alternative Agriculture* 1: 30-38.

Amanor, K.S. 1995. Dynamics of herd structures and herding strategies in West Africa: a study of market integration and ecological adaptation. *Africa* 65 (3): 351-394.

Amoah, F.E.K. 1976. Evolution of urban settlements on the Accra plains: early times to 1700 AD. *Bulletin of the Ghana Geographical Association* 18: 1-14.

Anchirinah, V., Ojiha, D., Owusu-Sekyere, R., Ramnanan, N. and Zhou, S.K. 1996. Production and Marketing of Yams in the Forest/Savanna Transition Zone of Ghana. *International Centre for Development Oriented Research in Agriculture,* Wageningen.

Anderson, D.J. 1986. Ecological Succession. In Kikkawa, J. and Anderson, D.J. 1986. *Community Ecology: Pattern and Process, 269-285.* Blackwell Scientific Publications, Melbourne.

Angstreich, M. 1991. Are trees effective against desertification? experiences from Niger and Mali. In Baxter, P.W.T. When the Grass is Gone: *Development Intervention in African Arid Lands - Seminar Proceedings,* 14 - 151, Scandinavian Institute of African Studies Uppsala.

Anquandah, J. 1982. Rediscovering Ghana's Past. Longman, Harlow.

Apffel Marglin, F. and Marglin, S.A. 1996. (Eds.) *Decolonising Knowledge: From Development to Dialogue.* Oxford University Press, New York.

Apostolopoulou, E. and Pantis, J.D. 2010. Development plans versus conservation: explanation of emergent conflicts and state political handling. *Environment and Planning A.* 42: 982-1000

Armar-Klemesu, M., Rikimaru, T., Kennedy, D.O., Harrison, E., Kido, Y., E.E.K. 1995. Household food security, food consumption patterns, and the quality of children's diet in a rural northern Ghana community. *Food and Nutrition Bulletin* 16 (1): 27-33.

Aryeetey, E. 1994. *Financial Integration and Development in Sub-Saharan Africa: a Study of Informal Finance in Ghana.* Overseas Development Institute, London.

Aryeety, E. and Steel W.F. 1995. Savings collectors and financial intermediation in Ghana. *Savings and Development* 19 (2):191-212.

Aryeety, E. and Udry, C. 1997. The characteristics of informal financial markets in sub-Saharan Africa. *Journal of African Economics* 6: 161-203.

Asibey, E.O.A. 1965. Utilisation of wildlife in Ghana. *The Ghana Farmer IX*(3): 91-93.

Asomaning, V., Grieco, M., Dankwa, Y., Apt, N., Grieco, M. and Turner, J. 1996. Stepping in and losing out: bread winning and educational loss amongst Ghana's school age girls. In Grieco, M and Apt, N. 1996. *At Christmas and on Rainy Days: Transport, Travel and the Female Traders of Accra,* 153-183. Avebury, Aldershot.

Asthana, R. 1996. Involuntary resettlement. survey of international experience. *Economic and Political Weekly* 31(24): 1468-1475.

Asuming-Brempong, S. and Breth, S.A. 1994. Effects of exchange rate liberalisation and input-subsidy removal on the competitiveness of cereals in Ghana. *Issues in African Rural Development* 2: 43-59.

Aubynn, A.K. 1996. Sapping the environment? *The case for structural adjustment programme (SAP) and its environmental implications for Ghana.* University of Helsinki, Helsinki.

Austin, G. 2008. Resources, techniques, and strategies south of the Sahara: revising the factor endowments perspective on African economic development, 1500-2000. *Economic History Review,* 61(3): 587- 624.

Awoonor, K.N. 1990. *Ghana: a Political History from Pre-European Times to Modern Times.* Sedco Publishing Ltd., Accra.

Awumbila, M., Momsen, J.H and Brookfield, H. 1995. Gender and the environment. *Global Environmental Change* 5(4): 337-346.

Ayoade, J.O. and Akintola, F.O. 1982. A note on some rainfall charactertics of rainstorms in Ibadan, Nigeria. *Weather* 37(2): 56 - 58.

Baidu Forson, J. 1997. On-station farmer participatory varietal evaluation: a strategy for client-oriented breeding. *Experimental Agriculture* 33 (1): 43-50.

Bailly, C., Barbier, C., Clement, S., Goudet, J.P., Hamel, O. 1982. Les problemes de la satisfaction des besoins en Afrique tropicale seche. Connaissances et incertitudes. Bois For Trop, 197 (23 - 43) Cited in Pieri, C.J.M.G. *Fertility of Soils: A Future for Farming in the West African Savannah,* 92. Springer Verlag, Berlin.

Baker, P.R. 1981. Environmental degradation in Kenya: social crisis or environmental crisis ? Paper for Developing Areas, Study Group of the Institute of British Geographers Conference, Leicester. Cited in Blaikie, P. *The Political Economy of Soil Erosion in Developing Countries.* Longman Group Ltd, Harlow.

Baker, R. 1986. Institutional innovation, development and environment: an administrative trap revisited. *Public Administration and Development* 9 (1): 32 - 47.

Bakkes, J. and Woerden, J.Van. 1997. (eds) *The Future of the Global Environment: a Model-Based Analysis Supporting UNEP's First Global Environment Outlook.* UNEP, Nairobi.

Ban, A.W. Van Den and Hawkins, H.S. 1996. *Agricultural Extension Blackwell Science,* Oxford.

Banerjee, D. and Mayerfeld Bell, M. 2007. Ecogender: locating gender in the environmental social sciences. *Society and Natural Resources.* 20(1): 3-19.

Banuri, T. 1990. Modernisation and its discontents: a cultural perspective on the theories of development. In Marglin, F.A. and Marglin, S.A. 1990. Dominating Knowledge: Development, *Culture and Resistance,* 73-101. Clarendon Press, Oxford.

Barbier, E. 2000. The economic linkages between rural poverty and land degradation: some evidence from Africa. *Agriculture, Ecosystems and Environment.* 82: 1-3.

Barrett, C.B. 1997. How credible are estimates of peasant allocative, scale, or scope efficiency? a commentary. *Journal of International Development* 9 (2): 221-229.

Barrow, C.J. 1991. *Land Degradation:Development and Break down of Terrestrial Environments.* Cambridge University Press, Cambridge.

Barry, R.G. and Chorley, R.J. 1982. *Atmosphere, Weather and Climate.* Methuen, London.

Barth, F. 1963. *The Role of the Entrepreneur in Social Change in Northern Norway.* Universitetforlaget, Bergen.

Bassett, T. and Crummey, D. (Eds.) 1993. *Land in African Agrarian Systems.* University of Wisconsin Press, Madison.

Batterbury, S.P.J. 1996. Planners or performers? Reflections on indigenous dryland farming in northern Burkina Faso. *Agriculture and Human Values.* 13 (3): 12-22.

Bauer, E. and Hoffman, V. 1997. Participatory rural appraisal: challenges and recommendations. *Quarterly Journal of International Agriculture* 36 (1): 81-99.

Bebbington, A. 1991. Indigenous agricultural knowledge systems, human interests, and critical analysis: reflections on farmer organization in Ecuador. *Agriculture and Human Values* 8 (1-2): 14-24.

Bebbington, A. 1996. *Movements, modernizations, and markets: indigenous organisations and agrarian strategies in Ecuador,* In Peet, R. and Watts, M.(eds) 1996. Liberation Ecologies: Development, Social Movements, 86-109. Routledge, London.

Becko, C. 1992. Studies of traditional conservation of biodiversity in Ghana. In Laing, E. (Ed.) 1992. *Proceedings of the Workshop on Biodiversity,* 32-33. Environmental Protection Council, Accra.

Bediako, J. 2004. *The Ghanaian Chronicle* 12(83): 1 -2.

Beeby, A. 1993. *Applying Ecology.* Chapman and Hall, London.

Begon, M., Harper, J.L. and Townsend, C.R. 1996. *Ecology: Individuals, Populations and Communities.* (3rd ed.). Blackwell Science Ltd., Oxford.

Belem, P.C. 1985. *Coton et systemes de production dans l'ouest du Burkina Faso.* University Paul Valery, Montpellier III. In Pieri, C.J.M.G. Fertility of Soils: A Future for Farming in the West African Savannah., 86, Springer - Verlag, Berlin.

Belsky, A.J. 1990. Tree/grass ratios in East African Savannas: a comparison of existing models. *Journal of Biogeography* 17: 483-489.

Bendini, M. and Palomares, M. 1995. Women's labour in the fruit growing industry: an approach to a conceptual model. *Paraguaya de Sociologia* 94:75-83.

Bendon, T. and Redclift, M. 1994. Introduction. In Redclift, M. and Benton, T. *Social Theory and the Global Environment*, 1-27. Routledge, London.

Benneh, G. 1985. Population, disease and rural development programmes in the Upper East Region of Ghana. In Clarke, J.L. *Population and Development Programmes in Africa,* 112 - 145, Cambridge University Press, Cambridge.

Benneh, G. and Agyepong, G.A. 1990. *Land Degradation in Ghana. Commonwealth Secretariat,* London.

Biersack, A. and Greenberg, J. B. 2006. *Reimagining Political Ecology: Culture/Power/History/Nature.* Duke University Press, Copenhagen.

Binns, T. 1994. *Tropical Africa Routledge,* London.

Birgegard, L.E. 1993. *Natural resource tenure: a review of issues and experiences with emphasis on sub-Saharan Africa*. Rural Development Studies No. 31. Swedish University of Agricultural Sciences, Sweden.

Black, R. and Girod, M. 1970. Late palaezoic to recent igneous activity in West Africa and its relationship to basement structure. In Clifford, T.N. and Gass, I.G. (Eds.) *African Magmatism and Tectonics,* 185 - 210, Oliver Boyd, Edinburgh.

Blackden, M and Wodon, Q. eds. 2006. *Gender, Time Use, and Poverty in Sub-Saharan Africa,* Washington DC: World Bank.

Blaikie, P. 1985. *The Political Economy of Soil Erosion in Developing Countries.* Longman, London.

Blaikie, P. and Brookfield, H. 1987. *Land Degradation and Society.* Methuen, London.

Blaikie, P. 1999. A review of political ecology: issues, epistemology and analytical narratives. *Zeitschrift fur Wirtschaftsgeographie* 43 (3–4): 131–147.

Bliss, F. 1996. *Women and Trees: Livelihood and Resource Management in Developing Countries* Beitrage zur Kulturkunde-Politischer Arbeitskreis Schulen.

Boateng, I.K. 1986. *Governmental and voluntary participation in Ghana's rural development programme.* In Brown, C.K. Rural Development in Ghana, 33 - 43. Ghana Universities Press, Accra.

Bond, W.J., Cowling, R.M. and Richards, M.B. 1992. Competition and coexistence. In Cowling, R.M. 1992. *The Ecology of Fynbos: Nutrients, Fire and Diversity, 206*-225. Oxford University Press, Cape Town.

Bonsu, M. 1998. *A study of vulnerability of cereal production to climate change and adaptation strategies in Ghana*. EPA, Accra.

Booth, A.H. 1959. On the mammalian fauna of the Accra Plain. *Journal of the West African Science Association* 5(1): 26-35.

Booth, D. 1993. Development research: from impasse to a new agenda. In Schuurman F.J. 1993. *Beyond the Impasse: New Directions in Development Theory*, 49-76. Zed Books, London.

Borlaug, N.E. and Dowswell, C.R. 1995. *Mobilising science and Technology to Get Agriculture Moving in Africa in Development Policy Review* 1995 13 (2): 115-129.

Bortei-Doku, E. 1967. *The need for efficient extension services.* Legon Agricultural News 1(5): 3-5.

Bortei-Doku, E. 1970. *Towards a More Efficient Extension Service.* Legon Agricultural News 4(1): 3-5.

Boserup, E. 1965. *The conditions of agricultural growth: the economics of agrarian change under population pressure.* George Allen and Unwin, London.

Boserup, E. 1981. *Population and Technological Change: a Study of Long Term Trends.* University of Chicago Press, Chicago.

Boughton, D., Crawford, E., Howard, J., Oemke, J. and Staatz, J. 1995. A strategic approach to agricultural research program planning in Sub-Saharan Africa. *MSU International Development Working Papers* No.49. Michigan State University, Michigan.

Boulding, K. 1966. Discussions. In Fraser Darling, F., and Milton, J.P. (eds) *Future Environments of North America, 291 to 292.* Natural History Press, New York.

Bowlby, S.R. and Lowe, M.S. 1992. Environmental and Green Movements. In Mannion, A.M. and Bowlby, S.R. (eds) *Environmental Issues in the 1990s, 161 to 174.* John Wiley and Sons, Chichester.

Brady, N.C. 1984. The Nature and Properties Of Soils (9th ed). *Macmillan Publishing Company*, New York.

Braun, B. and Castree, N. (eds) 1998. *Remaking reality: nature at the millennium.* Routledge, London.

Brittain, V. 1994. *Africa, the Lost Continent.* New Statesman and Society. 8th April, 20 - 22.

Brock, K. 1994. *Facilitating Sustainability: the Role of a Participatory Methodology Unpublished M.Sc Thesis,* University of London.

Brouwer, J., and J.M. Powell. 1998. Increasing nutrient use efficiency in West-African Agriculture: the impact of micro-topography on nutrient leaching from cattle and sheep manure. *Agriculture, Ecosystem and Environment* 71: 229–239.

Brown C.K. 1986. (Ed.) *Rural Development in Ghana.* Ghana University Press, Accra.

Brown, C.K. 1986. Urban Bias and Rural Development In Ghana. In Brown, C.K. *Rural Development in Ghana, 201 - 219.* Ghana Universities Press, Accra.

Brown, L.R. and Wolf E.C. 1984. *Soil Erosion: Quiet Crisis in the World Economy.* World Watch Institute, Washinton D.C.

Bruce, J.W. and Migot-Adholla, S.E. 1994. *Searching for Land Tenure Security in Africa.* Kendall/Hunt Publishing Company, Dubuque.

Bruegge, E.Vor Der., Plas, M., Dunford, C. and Stack, K.E. 1995. Credit with education: a self financing way to empower women. *Convergence* 28(3):26-35.

Brundtland, G.H. 1989. *Economica Ecologica.* El Pais Temas de nuestra epoca, 30 March p.4.

Bryant, R.L. 1992. Political ecology: an emerging research agenda in Third World studies. *Political Geography* 11 (1): 12-36.

Bryant, R.L. and Bailey, S. 1997. *Third World Political Ecology.* Routledge, London.

Bryant, R.L. and Goodman, M.K. 2004. Consuming narratives: the political ecology of 'alternative' consumption. *Transactions of the Institute of British Geographers,* 29(3): 344-366.

Bryceson, D.F. and Howe, J. 1993. Rural household transport in Africa: reducing the burden on women? *World Development* 21(11): 1715-1728.

Buckingham-Hatfield, S. 2000. *Gender and Environment.* Routledge, London.

Buckman, H.O. and Brady, N.C. 1969. *The Nature and Properties of Soils.* (7th ed.) Macmillan, London.

Budget-Meakin, C. 1995. Technology, Women and the Environment. Science, *Technology and Development* 13(1):72-78.

Bunyard, P. and Morgan-Grenville, F. 1987. *The Green Alternative.* Methuen, London.

Buringh, P. 1968. Introduction to the Study of Soils In *Tropical and Subtropical Regions.* Centre for Agricultural Publishing and Documentation, Wageningen.

Buringh, P. 1981. An Assessment of Losses and Degradation of Productive Agricultural Land in the World. *Food and Agricultural Organisation,* Rome.

Burke, T. 1984. The Politics of Ecology. In Hillary, E. 1984. Ecology 2000: *The Changing Face of the Earth,* 189-203, Multimedia Publications, London.

Burrows, C.J. 1990. *Processes of Vegetation Change.* Unwin Hyman, London.

Buttel, F.H. 1992. Environmentalisation: origins, processes, and implications for rural social change. *Rural Sociology* 57: 1-27.

Buttel, F.H. and McMichael, P., 1994. Reconsidering the explanation and scope of development studies: toward a comparative sociology of state-economy relations. In Booth, D. 1994. *Rethinking Social Development - Theory, Research and Practice,* 42-61. Longman, Harlow.

Byers, H.R. 1959. General Meteorology. McGraw Hill, New York.

Campbell, M. 1998. Interactions between biogeography and rural livelihoods in the coastal savanna of Ghana. *PhD Thesis,* University of London, London.

Campbell, M. and Palmer-Jones, R.W. 1999. *The utility and problems of GIS methods in the analysis of social imprints on the landscape of the central savanna of Ghana: A report to ESCOR.* University of East Anglia, Norwich.

Campbell, M. 2004. *The role of socio-environmental networking in the sustainability of rainfed agriculture in the coastal savanna of Ghana 61(1):* 79 -88.

Campbell, M 2005a. Actor networking, technological planning conceptions of space: the dynamics of irrigation farming in the coastal savanna of Ghana. *Applied Geography* 25: 367 – 381.

Campbell, M. 2005b. The ecological and social context of mammal hunting in the coastal savanna of Ghana. *Geoforum* 36: 667-680.

Carney, J. 1991. Indigenous soil and water management in Senegambian rice farming systems. *Agriculture and Human Values* 8 (1-2), 37-48.

Carr, M., Dignard, L. and Havet, J. 1995. Women, technology and small enterprise development. In Dignard, L. *Women in Micro and Small Scale Enterprise Development,* 217-227. Westview Press, Boulder.

Carson, W.P. 1985. *State of Knowledge Report: The Ecology of the Accra-Winneba Plains with Some Aspects of Related Savanna Ecosystems.* Environmental Protection Council, Accra.

Carson, W.P. 1985. *The ecology of the Accra-Winneba plains with some aspects of related savanna ecosystems.* Environmental Protection Council, Accra.

Carter, M.R. 1997. Environment, technology, and the social articulation of risk in West African agriculture. *Economic Development and Cultural Change* 45 (3): 557-590.

Casenave, A. and Valentin, C. 1992. A runoff capability classification system based on surface features criteria in semiarid areas of West Africa. *Journal of Hydrology* 13 (1): 231 - 249.

Cassaday, K.A., Monnet, G. and Dowswell, C.R. 1995. Strengthening national extension services in sub-Saharan Africa. SAA/Global 2000/CASIN. *Proceedings of a seminar for extension practioners,* 15-19 May 1994, Chavannes de Bogis. Centre for Applied Studies in International Negotiations (CASIN), Geneva.

Chambers, C. and Conway, G.R. 1992. *Sustainable rural livelihoods: practical concepts for the 21st Century.* Institute of Development Studies Discussion Paper 296 I.D.S. Publications, Brighton.

Chambers, R 1994a. The origin and practice of participatory rural appraisal. *World Development* 22(7): 953-969.

Chambers, R. 1994b. Participatory rural appraisal: analysis of experience *World Development* 22(9): 1253-1268.

Chambers, R., Longhurst, R. and Pacey, A. (Eds.) 1981. *Seasonal Dimensions to Rural Poverty* Frances Pinter, London.

Channa, J.A. 1968. *An Analysis of Thunderstorms Occurring in Namulonge, Uganda.* 4th Specialist Commitee on Applied Meterology, Nairobi.

Chen, M.A. 1995. *Beyond Credit: a Subsector Approach to Promoting Women's Enterprises*. Aga Khan Foundation, Ottawa.

Choin, G. 2002. Sacred groves in history: pathways to the social shaping of forest landscapes in coastal Ghana. *IDS Bulletin* 33(1): 39 – 46.

Chopra, K. 1996. The management of degraded land: issues and an analysis of technological and institutional solutions. *Indian Journal of Agricultural Economics* 51 (1-2): 238-248.

Christiansen, N.L. 1993. Fire regimes and ecosystem dynamics. In Crutzen, P. and Goldhammer 1993. *Fire in the Environment: The Ecological, Atmosphere, and Climatic Importance of Vegetation Fires.* John Wiley and Sons, Chichester.

Christoplos, I. 1996. Poverty, pluralism and extension practice. *Gatekeeper Series- International Institue for Environment and Development* No.64 Uppsala, Sweden.

Church, R.J.H. 1980. *West Africa.* Longman Group Ltd., London.

CIA 2012. *The world factbook.* URL: http://wwwciagov/cia/ publications/ factbook/

Ciancaglini, N.C. 1996. Social factors influencing the selection and implementation of mechanised irrigation systems *ICID Journal* 45(2): 21-29.

Clark, N., Perez-Trejo, F. and Allen, P. 1995. *Evolutionary Dynamics and Sustainable Development: A Systems Approach* Edward Elgar, Aldershot.

Clay, Daniel C., and L. A. Lewis. 1990. Land use, soil loss and sustainable agriculture in Rwanda. *Human Ecology* 18(2):147-161.

Clifford, T.N. 1970. The Structured Framework of Africa. In Clifford, T.N. and Gass, I.G. *African Magmatism and Tectonics* (Eds.) 1 -26, Oliver Boyd, Edinburgh.

Cline-Cole, R.A. 1995. Livelihood, sustainable development and indigenous forestry in dryland Nigeria. In Binns, T. (Ed.) *People and Environment in Africa.* John Wiley and Sons, Chichester.

Cole, M.M. 1986. The savannas of West Africa and the South Western Sudan. In Cole, M.M. The Savannas, *Biogeography and Geobotany,* 264 - 290, Academic Press, London.

Collinet, J. and Valentin, C. 1984. Evaluation of factors influencing water erosion in West Africa using rainfall simulation. In Walling, D.E., Foster, S.S.D. and Wurzel, P. (Eds.) *Challenges in African Hydrology and Water Resources,* 451 - 461., I.A.H.S. Publication No. 144, New York.

Columbani, T., Olivry, J.C. and Kallel, R. 1984. Exceptional events in erosion and sediment transport in arid and semiarid Africa. In Walling, D.E., Foster S.S.D. and Wurzel P. (Eds.) *Challenges in African Hydrology and Water Resources,* 295 - 500. I.H.A.S. Publication No. 144, New York.

Common, M. 1988. *Environmental and Resource Economics: An Introduction.* Longman London.

Common, M. and Stagl, S. 2005. *Ecological Economics. An Introduction.* Cambridge University Press.

Convery, F.J. 1995. *Applying Environmental Economics in Africa* World Bank Washington.

Conway, G.R. 1982a. *KKU-Ford Cropping Systems Project. An Agroecosystem Analysis of Northeast Thailand.* University of Khon Kaen, Khon Kaen.

Conway, G.R. 1982b. *Applying Ecology.* Inaugural Lecture, ICCET, Imperial College, London.

Conway, G. 1983a. *Agroecosystem Analysis.* I.C.C.E.T. Series E, Imperial College, London.

Conway, G.1983b. *Applying Ecology.* Centre for Environmental Technology, Imperial College, London.

Conway, G., 1985a. Agricultural Ecology and Farming Systems Research. In Remenyi, J.R. (Ed.) *Agricultural Systems Research for Developing Countries.* Australian Centre for International Agricultural Research, Canberra.

Conway, G. 1985b. *Agroecosystems Analysis.* Agricultural Administration, 20: 31 - 35.

Conway G.R. 1987. *The Properties of Agroecosystems.* Agricultural Systems 24: 95-117.

Copans, J. 1975. *Quile Nourrit de la Famine en Afrique?* and Secheresses et Famines du Sahel, F. Maspero, Paris.

Cotthem, W van, Beel, C., Danneels, P., Keyer, J. de., Vyvey, Q., Van Cotthem, W., De-Keyzer, J. 1991. Restoring the natural vegetation on strong eroded volcanic soils of Cape Verde (West Africa). *Soil Technology* 4 (1): 183 - 189.

Coulter, J.K. 1972. Soil management systems. In Coulter, J.K. *Soils of the Humid Tropics,* 189 - 197, National Academy of Sciences, Washington D.C.

Cowling, R.M. 1992. *The Ecology of Fynbos: Nutrients, Fire and Diversity* Oxford University Press, Cape Town.

Cowling, R.M. and Holmes, P.M. 1992. Flora and vegetation. In Cowling, R.M. 1992. *The Ecology of Fynbos: Nutrients, Fire and Diversity,* 23-61, Oxford University Press, Cape Town.

Cowling, R.M., Holmes, P.M. and Rebelo, A.G. 1992. Plant diversity and endemism. In Cowling, R.M. 1992. *The Ecology of Fynbos: Nutrients, Fire and Diversity,* 62-112, Oxford University Press, Cape Town.

Craib, I. 1992. *Anthony Giddens Routledge,* London.

Crasswell, E.T. 1993. The Management of World Soil Resources for Sustainable Agricultural Production. In Pimentel, D. (Ed.) *World Soil Erosion and Conservation,* 257 - 275. Cambridge University Press, Cambridge.

Crehan, K. 1992. The household as a factor in farmer decision making. In Berstein, H., Crow, B. and Johnson H. (Eds.) *Rural Livelihoods: Crises and Responses, 113*-135. Oxford University Press, Oxford.

Crow, B. 1992. Rural livelihoods: action from above. In Berstein, H., Crow, B. and Johnson H. (Eds.) *Rural Livelihoods: Crises and Responses,* 251-271, Oxford University Press, Oxford.

Daly, H. 1980. *Economics, Ecology and Ethics.* Freeman, San Francisco.

Dankelman, I. and Davidson, J. 1988. *Women and Environment in the Third World: Alliance for the Future.* Earthscan Publications Ltd. London.

Dankelman, I. 2010. *Gender and climate change: an introduction.* Earthscan, Oxford.

Danquah, F. 1995. *Cocoa Diseases and Politics in Ghana 1909-1996.* Peter Lang, New York

Darkoh, M.B.K. 1996. The Human dimension of desertification in the drylands of Africa. *Journal of Social Development in Africa* 11(2): 89-106.

Darwin, C. 1859. *The Origin of Species by Means of Natural Selection.* John Murray, London.

Davies, S. 1992. Green conditionality and food security: winners and losers from the greening of aid. *Journal of International Development* 4(2): 151 to 165.

Day, J.C. and Aillery, M.P. 1988. Soil and moisture management in Mali: a case study analysis for West Africa. *Agricultural Economics.* 2 (3): 209 - 222.

De Miranda, E.E., 1977. *Rapport de Mission d' Etude Agronomique Sur les Villages temoins,* Dan Koulou, Gourjae, Magami et Serkin Haoussa. Univ. Bordeaux 11 - DG RST-IRAM.

Deacon, H.J., Jury, M.R. and Ellis, F. 1992. Selective regime and time. In Cowling, R.M. (Ed.) *The Ecology of Fynbos: Nutrients, Fire and Diversity,* 6-22, Oxford University Press, Cape Town.

Demery, L. and Squire, l. 1996. Macroeconomic adjustment and poverty in Africa: an emerging picture. World Bank Research. *Observer* 11 (1):39-59.

Denevan, W. 1985. *Terrace Abandonment in the Peruvian Andes: Extent Causes and Prospects for Restoration.* Proc. Consejo Nacional de cienda y Technologica Seminian - Taller, Lima.

Denevan, W. 1985. *Terrace Abandonment in the Peruvian Andes: Extent, Causes and Prospects for Restoration.* Proc. Consejo Nacional de cienda y Technologica Seminian - Taller, Lima.

Dennis, C. and Gargett, M. 1995. *Issues in the Planning of Community Based Projects and Programmes: Development Projects: Issues for the 1990.* University of Bradford, Bradford.

Department of Parks and Gardens Report 1993. *Government of Ghana,* Accra.

Derrick, J. 1984. West Africa's worst year of famine. *African Affairs* 83(332): 281 - 299.

Diane, M.K. 1968. Rice Mechanisation. *Ninth Annual Agricultural Conference,* 102-106. U.S.A.I.D/Ghana, Accra.

Dickson K.B. and Benneh, G. 1988. *A New Geography of Ghana.* Longman Group Ltd., London.

Dickson, K.B. 1964. The agricultural landscape of southern Ghana and Ashanti-Brong Ahafo: 1800 to 1850. *Bulletin of the Ghana Geographical Association* 9(1): 25-35.

Dickson, K.B. 1966. Historical geography: some problems of presentation and interpretation, *Bulletin of the Ghana Geographical Association* 2: 5-14.

Dickson, K.B. and Benneh, G. 1988. *A New Geography of Ghana.* Longman, London.

Diham, B. and Hines, C. 1983. *Agribusiness in Africa.* Earth Resource Research Ltd. London.

Dixon, R.K., Perry, J.A., Vanderlein, E.L. and Hiol, F.H. 1996. Vulnerability of forest resources to global climate change: case study of Cameroon and Ghana. *Climate-Research* 6 (2):127-133.

Diykman, J.T., Mallinkrodt-Neidhandt, J., and Boers, T.M. 1990. The threat of erosion. *Development and Cooperation.* 4 (1): 20 - 26.

Dobson, A. 1990. *Green Political Thought.* Unwin Hyman, London.

Dodoo, F.N. and Landewijk, P.Van. 1996. Men, women, and the fertility question in sub-Saharan Africa: an example from Ghana. *African Studies Review* 39(3):29-41.

Dorm-Adzobu, C. 1995. *New Roots: Institutionalizing Environmental Managememt in Africa* World Resources Institute, Washington D.C.

Dove, M.R., and Carpenter, C. (Eds). 2008. *Environmental Anthropology: A Historical Reader.* Blackwell, MA.

Driel, W.F. van, and Van - Driel, W.F. 1990. Is sustainability widely accepted? A critical review of soil and water conservation projects in the Sahel. *AT - Source* 18 (3), 27 - 31.

Dudal, R. 1981. An Evaluation of Conservation Needs. In Morgan, R.P.C. (Ed.) *Soil Conservation, Problems and Prospects,* 3 -12. John Wiley and Sons, Chichester.

Dunn, J. and Robertson, A.F. 1973. *Dependence and Opportunity: Political Change in Ahafo.* Cambridge University Press, Cambridge.

Dupuis, J. 1824. Journal of a Residence in Ashantee, London. Cited in Dickson, K.B. 1964. *The Agricultural Landscape of Southern Ghana and Ashanti-Brong Ahafo: 1800 to 1850 Bulletin of the Ghana Geographical Association* 9(1) 25.

Eckholm, E.P. 1976. *Losing Ground.* Norton, New York.

Eckolm, E.P. 1978. *Losing Ground: Environmental Stress and World Food Production.* Pergamon Press, London.

Eden, M.J. and Parry, J.T. 1996. *Land Degradation in the Tropics: Environmental and Policy Issues.* Pinter, London.

Edwards P.J., May, R.M. and Webb, N.B. 1994. *Large Scale Ecology and Conservation Biology.* Blackwell Scientific Publications, Oxford.

Eger, H., Fleischhauer, E., Hebel, A., and Sombroek, G. 1996. Taking action for sustainable land use: results from the 9th ISCO Conference in Bonn, Germany. *Ambio* 25 (8): 480-483.

Ehlrich, P. and Raven, P.H. 1964. Butterflies and plants: a study in coevolution. *Evolution* 18: 586-608.

Ehrlich, P.R., 1968. *The Population Bomb.* Ballantine, New York.

El - Mourisi, A.W.A. 1984. The role of higher agricultural education in the improvement of shifting cultivation farming systems in Africa. In Bunting, E. The Future of Systems in Africa and the Task of Universites: Proceedings of the International Workshop on Shifting Cultivation *Teaching and Research at the University Level.* 4th to 9th July 1992, 8 - 14, Fao, Rome.

El - Swaify, S.A. 1993. Soil erosion and conservation in the humid tropics. In Pimentel, D. *World Soil Erosion and Conservation,* 233 - 255. Cambridge University Press, Cambridge.

El - Swaify, S.A., Dangler. E.W. and Armstrong, C.L. 1982. Soil erosion in the tropics. *University of Hawaii Research Extension Series* 24: 173.

Ellen, R. 1996. Introduction. In Ellen, R. and Fukui, K. (eds.) 1996. *Redefining Nature, Ecology, Culture and Domestication,* 1-36. Berg, Oxford.

Ellery, W.N., Mentis, M.T. and Scholes, R.J. 1992. Modelling the location of woody-grassland boundaries. In Furley, P.A., Proctor, J. and Ratter, J.A. 1992. *Nature and Dynamics of Forest-Savanna Boudaries,* 569-582, Chapman and Hall, London.

Greenberg, J. B. and Park, T.K. 1994. Political Ecology. *Journal of Political Ecology* 1: 1-12.

Elliot, J. 1994. *An Introduction to Sustainable Development in the Developing World.* Routledge, London.

Elliot, J.A. 1995. *Government policies and the population environment interface: land reform and distribution in Zimbabwe.* In Binns, T. 1995. People and Environment in Africa. John Wiley and Sons, Chichester.

Elliott, J. A. and Campbell, M. 2002. The environmental imprints and complexes of social dynamics in rural Africa: cases from Zimbabwe and Ghana. *Geoforum* 33: 221–37.

Ellis, F. 1988. *Peasant Economics: Farm Households and Agrarian Development.* Cambridge University Press, Cambridge.

Ellis, F. 1993. P*easant Economics: Farm Households and Agrarian Development* (2nd. ed.). Cambridge University Press, Cambridge.

Elwood, S. 2010. Mixed Methods: Thinking, Doing, and Asking in Multiple Ways. In Delyser, D., Herbert, S., Aitken, S., Crang, M. and McDowell, L. (Eds.) *Qualitative Geography,* 94 - 113. Sage, London.

Endeley, J., Tetebo, R. and Breth, S.A. 1996. Gender Considerations in Designing Agricultural Extension Programs in Achieving Greater Impact from Research Investments in Africa. *Proceeding of the Workshop Developing African Agriculture: Achieving Greater Impact from Research Investment.* Addis Ababa, Ethiopia 26-30 September, 1995, 150-165. Sasakawa Africa Association c/o CIMMYT, Mexico City.

Engberg, L.B. 1972. An extension problem: understanding the development process. *Legon Agricultural News* 5(2): 12-17.

Entsua-Mensah, C. 1995. Linking research and extension services for sustainable food production in Ghana: some options for achievement. *Quarterly Bulletin of IAALD* 40(4):121-125.

Environmental Management Associates 1989. *Environmental Study of Accra Metropolitan Area Final Report.* Government of Ghana, Accra.

Eponou, T. 1996. *Partners in Technology Generation and Transfer: Linkages Between Research and Farmer's Organisations in Three Selected African Countries. International Service for National Agricultural Research,* The Hague.

Ervin, A.M. 2005. *Applied Anthropology: Tools and Perspectives for Contemporary Practice.* Pearson, Boston.

Escobar, A. 1995. *Encountering development: the making and unmaking of the Third World.* Princeton University Press, Princeton.

Escobar, A. 1996. Constructing nature: elements for a poststructural political ecology. In Peet, R. and Watts, M. 1996. *Liberation Ecologies: Environment, Development, Social Movements,* 46-68, Routledge, London.

Eweg, H.P.A. and Lammeren, R. Van. 1996. *The Application of a Geographical Information at the Rehabilitation of Degraded and Degrading Areas: a Case Study in the Highlands of Tigray,* Ethiopia. Agricultural University, Wageningen.

Ewusi, K. 1986. The Contribution of Research Institutions and the Universities to Rural Development in Ghana. In C.K. Brown (Ed.) *Rural Development in Ghana.,* 165 - 180. Ghana Universities Press, Accra.

Eyzaguirre, P. 1996. *Agricultural and Environmental Research in Small Countries: Innovative Approaches to Strategic Planning.* John Wiley and Sons, Chichester.

Faber, D. 1993. *Environment Under Fire: Imperialism and the Ecological Crisis in Central America.* Monthly Review Press, New York.

Fairhead, J. and Leach, M. 1994. *Contested forests: modern conservation and historical land use in Guinea's Ziama reserve. African Affairs* 93(373):481-513.

Fairhead, J. and Leach, M. 1995. False forest history, complicit social analysis: rethinking some West African environmental narratives. *World Development* 23(6): 1023-1035.

Fairhead, J. and Leach, M. 1996a. *Misreading the African Landscape: Society and Ecology in a Forest Savanna Mosaic.* Cambridge University Press, Cambridge.

Fairhead, J. and Leach, M. 1996b. *Local agro-ecological management and forest-savanna transitions: the case of Kissidougou, Guinea.* In Binns, T. 1996. People and Environment in Africa, 163-170. John Wiley and Sons Ltd, London.

Fairhead, J. and Leach, M. 1995. False forest history complicit social analysis: Rethinking some West African environmental narratives. *World Development* 23(6): 1023-035.

Fairhead, J. and Leach, M.1998. *Reforming deforestation: global analysis and local realities – studies in West Africa.* Routledge, London.

Fauck, R. 1977. Erosion and mechanisation. In Greenland, D.J. and Lal, R. *Soil Conservation and Management in the Humid Tropics 189 - 193.* John Wiley and Sons, Chichester.

Fenley, J. 1979. *The Equatorial Rain Forest: a Geological History.* Butterworths, London.

Fiege, K. 1995. *Farmers' Self-Help Schemes in the Management of Natural Resources in Southern Mali: Possibilities for Institutional Support in the Districts of Tominian and Bla.* Humboldt University, Berlin.

Findlay A.M. and Findlay, A.M. 1987. *Population and Development in the Third World.* Methuen, London.

Finger, M. and Verlaan, P. 1994. Learning our way out: a conceptual framework for social-environmental learning. *World Development* 23(3): 503-513.

Fischer, N.M. 1987. Technical and environmental limitations to crop production in sub-humid and semi-arid West Africa. In *FAO Improving Food Crop Production on Small Farms in Africa,* 78 - 85. F.A.O. Rome.

Floro, M.S. 1995. Economic restructuring, gender and the allocation of time. *World Development* 23(11): 1913-1929.

Folly, A. 1995. Estimation of erodibility in the savanna ecosystem, northern Ghana. Communications in *Soil Science and Plant Analysis.* 26(5-6): 799-812.

Food and Agricultural Organisation (FAO) 1996. *Food Security: A Domestic Approach FAO, Rome*

Food and Agricultural Organisation (FAO) 2003. Ghana. FAO, Rome.

FAO 2005. URL: http://www.fao.org/ag/agl/aglw/aquastat/countries/ghana/index.stm

Food and Agriculture Organization (FAO) 2010. *Gender dimensions of agricultural and rural employment: Differentiated pathways out of poverty. Status, trends and gaps*. FAO Economic and Social Development Department, Rome.

Fosu, K.Y. and Breth, S.A. 1994. Determinants and effectiveness of government expenditure policy in Ghana's agricultural sector. *Issues in African Rural Development* 2: 61-74.

Foth, H.D. and Schafer, J.W. 1980. *Soil Geography and Land Use.* John Wiley and Sons, New York.

Fourier, F. 1967. Research on Soil Erosion and Soil Conservation in Africa. *African Soil.* 22 (1): 53 - 96.

Franke, R. and Chasin, B.H. 1980. *Seeds of Famine: Ecological Destruction and the Development Dilemma in the Western Sahel.* Allenheld, Osmun and Company, New Jersey.

Freebain, D.K. 1995. Did the green revolution concentrate Incomes? A quantitative study of research reports. *World Development* 23, No. 2, 265-279.

Freud, E.H. and Freud, C. 1996. Devaluation and Liberalisation as Tools for Enhancing Competitiveness? *Some Insights from the Recent West African Experience in Cocoa Marketing.* CIRAD-SAR, Montpellier.

Friedman, J. 1992. *Empowerment: the Politics of Alternative Development.* Blackwell, Oxford.

Friends of the Earth 1992. *Plunder in Ghana's Rainforest for Illegal Profit: An Expose of Corruption, Fraud and Other Malpractices in the International Timber Trade.* Friends of the Earth, London.

Friends of the Earth 1996. *Ghana Fisheries-What Future. Foeline* 4: 8-9, FOE, Accra.

Frost, P.G.H., Menaut, J.C., Walker, B.H., Medina, E., Solbrig, O.T., Swift, M. 1986. Responses of savannas to stress and disturbance. In Scholes, R.J. and Walker, B.H. 1993. *An African Savanna: Synthesis of the Nylsvley Study,* 2-10. Cambridge University Press, Cambridge.

Fujisaka, S. 1997. Research: help or hindrance to good farmers in high risk systems? *Agricultural Systems* 54 (2): 137-152.

Furley, P., Proctor, J. and Ratter, J.A. 1992. (Eds) *Nature and Dynamics of Forest-Savanna Boundaries.* Chapman and Hall, London.

Gadgil, M. and Guha, R. 1992a. Interpreting Indian environmentalism. *Paper to the Conference on the Social Dimensions of Environment and Sustainable Development, United Nations Research Institute of Social Development,* Valetta, Malta, 22-25 April.

Gadgil, M. and Guha, R. 1992b. *This Fissured Land. An Ecological History of India.* Oxford University Press, Dehli.

Gakou, M. and Force, J.E. 1996. Learning with farmers for policy changes in natural resource management. In *Forests, Trees and People Newsletter* No.31, 14-19 Unite de Gestion Forestiere (UGF), Bamako.

Galan, B.B. 1994. *Farmer participation for a sustainable agriculture in Latin American countries.* Participation Popular No.7. FAO, Rome.

Gallopin, G. and Berrera, C. 1979. *The Nexus society and environment. In Environment and Styles of Development: Some Conceptual and Methodological Issues.* Technical Research Project No. 35, I.F.D.A.

Gayi, S.K. 1995. Adjusting to the social costs a adjustment in Ghana: problems and prospects. *European Journal of Development Research* 7 (1): 77-100.

Geertz, C. 1973. *The Interpretation of Cultures.* Basic Books, New York.

Gervy, R. 1987. The Fertilisation of some food crops in West Africa (Sudano-Sahelian Zone). In *Fertiliser Use and Crop Production.* (ed. F. Isherwood), 157 - 165. International Fertiliser Industry Association Ltd., Paris.

Governmennt of Ghana 1988. *Ghana Environmental Action Plan.* Ghana Publishing Corporation, Accra.

Ghana Ministry of Agriculture 2003. *Landuse, vegetation, Soils and Climate.* Government of Ghana, Accra.

Ghersi, G. and Martin, F. 1996. Towards a renewed and integral approach to food security. *Agroalimentaria* 2: 19-30.

Giddens, A. 1979. Central Problems in Social Theory. Macmillan, London.

Giddens, A. 1984. *The Constitution of Society.* Polity Press, Cambridge.

Gilbertson, D.D., Kent, M. and Pyatt, F.B. 1985. *Practical Ecology for Geography and Biology: Survey, Mapping and Data Analysis.* Hutchinson, London.

Gill, G.J. 1991. *Seasonality and Agriculture in the Developing World: A Problem of the Poor and Powerless.* Cambridge University Press, Cambridge.

Girvan, L.A. 1995. *Namibia: National Report on Women, Agriculture and Rural Development for the Fourth World Conference on Women.* Namibian Economic Policy Research Unit, Windhoek.

Glewwe, P. and Ilias N. 1996. The determinants of school attainment in sub-Saharan Africa: a case study of Ghana. *Journal of International Development* 8 (3): 395-413.

Godwin, P. 1986. Plague of Locust Puts Africa Under Threat. *The Times,* 24/8/86, 14.

Gold Coast Census Report-1948. *Colonial Service,* Accra.

Goldman, A. and Smith, J. 1995. Agricultural transformations in India and northern Nigeria: Exploring the nature of green revolutions, *World Development* 23(2): 243-263.

Goodland, R. 1984. *Environmental Mangement in Tropical Agriculture.* Westview Press, Boulder.

Goudie, A. 1980. (3rd ed.) *The Human Impact on the Natural Environment.* Basil Blackwell Ltd., Oxford,.

Gourou, P. 1966. The Tropical World. (4th ed). *Longmans Green and Company,* London.

Government of Ghana 2000. *Census Report.* Government of Ghana, Accra.

Graham, C.K. 1968. *A Report on the Cattle Industry Around Nungua.* Unpublished B.Sc. Dissertation, University of Ghana.

Grellier, R. 1995. *All In Good Time: Women's Agricultural Production in Sub-Saharan Africa.* University of East Anglia, Norwich.

Grigsby,W.J. 1996. Women, descent and tenure success among the Bambara of West Africa: a changing landscape. *Human Organisation* 55(1):93-98.

Grossman, L. 1981. The Cultural ecology of economic development. *Annals of the Association of American Geographers* 71 (2):220-236.

Grossman, L. 1984. *Peasants, Subsistence Ecology and Development in the Highlands of Papua New Guinea.* Princeton University Press, Princeton.

Guha, R. 1994. *The Environmentalism of the Poor presented at the Conference Disssent and Direct Action in the Late Twentieth Century.* Otavalo, Ecuador.

Gyasi, E., Agyepong, G.T., Ardayfio-Schandorf, E., Enu-Kwesi, L., Nabila, J.S. and Owusu-Bennoah E. 1995. Production pressure and environmental change in the forest-savanna zone of southern Ghana. *Global Environmental Change* 5(4):355-366.

Haakonsen, J.M. 1988. *Recent Developments of the Artisanal Fisheries in Ghana.* FAO, Cotonou.

Hakeem, S.A. and Vemuri, S.R. 1997. Structural adaption of human settlements to climate for sustainable agriculture. *Journal of Sustainable Agriculture* 10 (1):63-74.

Hall, D.O. and Long, S.P. 1993. Introduction: photosynthesis and the changing environment. In Hall, D.O., Scurlock, J.M.O., Bolhar-Nordenkampt, H.R., Leegood, R.C. and Long, S.P. (eds) 1993. *Photosynthesis and Production in a Changing Environment,* xxii., Chapman Hall, London.

Hall, J.B and Swaine, M.D. 1981. *Distribution and Ecology of Vascular Plant in a Tropical Forest.* DW Junk Publishers, The Netherlands.

Hall, J.B. 1985. *Personal communication in Carson* 1985. op cit p.45.

Hambly, H. 1996. Grassroots indicators: measuring and monitoring environmental change at the local Level. *ILEIA Newsletter* 12(3):14-15.

Handa, S. 1994. Gender, headship and intrahousehold resource allocation. *World Development* 22(10): 1535-1547.

Hanna, K. S., Clark, D.A. and Scott Slocombe, D. 2008. *Transforming Parks and Protected Areas: Policy and Governance in a Changing World.* Routledge, New York.

Harlan, J.R. 1977. The Plants and Animals that Nourish Man. In Scientific American Book, 57, *Food and Agriculture.* Freeman, San Francisco.

Harris, D.R. 1980. *Human Ecology in Savanna Environments Academic,* London.

Harrison Church, R. 1980. *West Africa* Longman, London.

Hatzius, T. 1996a. Sustainability and institutions: catchwords or new agenda for ecologically sound development? *Working Paper No.48.* Institute of Development Studies, Brighton.

Hatzius, T. 1996b. *The Institutional Dimension of Sustainability in Rural Development Diskussionsschriften* - Forschungsstelle fur Internationale Agrarentwicklung e.V.

Hawkins, R. 1994. *Use of Priority Setting and Causal Analysis Methods for Research and Development.* CIRAD-SAR, Montpellier.

Hawthorne, W.D. 1994. *Fire Damage and Forest Regeneration in Ghana: Forestry Inventory and Management Project of the Ghana Forestry Department.* Overseas Development Administration, London.

Hawthorne, W.D. and Abu-Juam, M. 1995. *Forest Project in Ghana: With Particular Reference to Vegetation and Plant Species* International Union of Nature and Natural Resources IUCN) Gland.

Heck, S. 1996. *Sales Contracts and Land Tenure Relations in Ankole,* Western Uganda. Boston University, Boston.

Hekstra, G.P. 1985. Development policies ignore ecological constraints. *The Ecologist* 15(5/6): 240-245.

Henderson -Sellers, A. and Robinson P.J. 1966. *Contemporary Climatology.* Longman Group UK Ltd., Harlow.

Henderson, H.K. and Hansen, E. 1995. *Gender and Agricultural Development: Surveying the Field* University of Arizona Press, Tuscon.

Herlihy, P.H. 1989. Panama's Quiet Revolution: Comarca Homelands and Indian Rights *Cultural Survival Quarterly* 13 (3): 17-24.

Hernandez, R. 1995. *Financial Management of Farm Households and Its Linkages with the Informal Financial Sector in Northern Ghana.* Institut fur Rurale Entwicklung, Gottingen.

Hernis, P.O. 1991. Modernizing Ghanaian Fisheries. *The Need for Social Carriers of Technology.* Ad Notham, Olso.

Hershkovitz, L. 1993. Political ecology and environmental management in the Loess Plateau, China, *Human Ecology* 21(4): 327-353.

Hiadzi, H.H.S. 1970. *Farm Mechanisation in the Greater Accra District* Unpublished B.Sc. Dissertation, University of Ghana.

Hickman, G.M. and Wright, D.R. 1974. *Ghana: Progress and Problems.* Common Ground, Harlow.

Higgins, G.M. and Kassam, A.H. and Naiken, L. 1982. *Potential Population Supporting Capacities of Land in the Developing World.* F.A.O., Rome.

Hjort Af Ornas, A. 1996. Approaching Nature from Local Communities: Security Perceived and Achieved. Linkoping University, Linkoping.

Hochet, A.M.and Aliba, N. 1995. *Rural Development and Participatory Methods in Africa: Research-Action-Development, a Way of Listening,* Commitment and Practice L'Harmattan, Paris.

Holling, C.S. 1973. Resilience and stability of ecological systems. *Annual Review of Ecological System* 4: 1-24.

Holling, C.S. 1993. *Investing in Research for Sustainability Ecological Applications* 3: 552-555.

Hoogmoed, W.B. 1986. Crust and sealing problems on West African soils. In Callebaut, F., Grabriels D. and de Boodt, M. (Eds.) *Assessment of Soils Surface Sealing and Crusting, 48 - 55.* Flanders Research Centre for Soil Erosion and Soil Conservation, Brussels.

Hope, K.B. 1984. *The Dynamics of Development and Development Administration.* Greenword Press, Westport.

Hopkins, B. 1992. Ecological processes at the forest savanna boundary. In Furley, P.A., Proctor, J. and Ratter, J.A. 1992. *Nature and Dynamics of Forest-Savanna Boundaries,* 21-33. Chapman and Hall, London.

House-Midamba, B. and Ekechi F.K. 1995. *African Market Women and Economic Power: the Role of Women in African Economic Development.* Greenwood Press Incorporated, Westport.

Howard, R. 1978. *Colonialism and Underdevelopment in Ghana.* Croon Helm, London.

Hudson, N.W. 1983. Soil Conservation Strategies in the Third World. *Journal of Soil and Water Conservation* 1983 38(6):446-450.

Hudson, N.W. 1977. The Factors determining the extent of soil erosion. In Greenland, D.J. and Lal, R. (Eds.) *Soil Conservation And Management in the Humid Tropics,* 11 - 16. John Wiley and Sons, Chichester.

Hulme, M. 2001. Climate perspectives in sahelian desiccation. *Global Environmental Change* 11: 19–29.

Hutchinson, R.A. 1962. Stock and Methods of Animals Husbandry. In Wills, J.B. 1962. *Agriculture and Land Use in Ghana,* Oxford University Press, London.

Hutton, W. 1821. *A Voyage to Africa in to Africa in the Year 1820.* Colonial Office London.

Iddrisu, A. and Tropp, S. 1995. Land Degradation and Desertification in Ghana. In Atchia, M. 1995 (Ed.) *Environmental Management: Issues and Solutions,* 125-127 John Wiley and Sons, Chichester.

Infoscience 1994. *Report on Sustainable Development in the Sesfontein*/Khowarib Basin Unpublished Report: Windhoek.

Ingham, B. 1993. The Meaning of development: interactions between "new" and "old" ideas. *World Development* 21(11): 1803 to 1823.

International Research Institute 1995. *A 2020 Vision for Food, Agriculture and the Environment: the Vision, Challenge, and Recommended Action.* In Workshop: Policy Implications of Global Trends in Food Production: Reader. International Research Institute, Washington DC.

International Food Policy Research Institue (IFPRI) 2003. *IFPRI Forum* URL:http://www.ifpri.org/sites/default/files/pubs/pubs/newsletters/ifpriforum/if3.pdf

International Fund for Agricultural Development (IFAD) 1988. Report of the Special Programming Mission to Ghana July 1988. In UNICEF 1990) *Children and Women of Ghana: A Situation Analysis* p.205, UNICEF, Accra.

International Labour Organization (ILO) 1995. *Action to Assist Rural Women: Lessons from Africa.* Case Studies of Four Projects in Zimbabwe, Tanzania, Niger and Guinea. ILO, Geneva.

International Union for the Conservation of Nature 1980. *World Conservation Strategy*. I.U.C.N./W.W.F./U.N.E.P, Gland.

Irrigation Development Authority 2000. IDA, Accra.

Irrigation Development Authority 1994. I.D.A., Accra.

Irvine, F.R. 1947. *The Fish and Fisheries of the Gold Coast*. Longman, London.

Irvine, S. and Ponton, A. 1988. *A Green Manifesto: Policies for a Green Future. Macdonald and Co.*(Publishers) Ltd, London.

Ite, U.E. 1997. Small Farmers and Forest Loss in Cross River National Park, Nigeria *The Geographical Journal* 163(1): 47-56.

Jacobs, M. 1991. The Green Economy, Environment, *Sustainable Development and Politics of the Future*. Pluto Press London.

Jackson, C. 1993. Doing what comes naturally? Women and environment in development. *World Development* 21(12): 1947-1963.

Jackson, C. 1994. Gender analysis and environmentalisms. In Redclift, M. and Benton, T. 1994. *Social Theory and the Global Environment,* 113-149. Routledge, London.

Jackson, I.J. 1977. *Climate, Water and Agriculture in the Tropics*. Longman Group Ltd., Harlow.

Jacobs, S. 1996. Structures and processes: land, families and gender relations. *Gender and Development* 4 (2):35-42.

Jaffee, S. and Morton, J. 1995. *Marketing Africa's High Value foods: Comparative Experiences of An Emergent Private Sector*. Kendall/Hunt Publishing Company, Dubuque.

Jago, N. 1987. The Return of the Eighth Plague. *New Scientist* 114 (1565), 47 - 51.

Jampoh, E.L. 1989. The Organisation of Fuelwood and Charcoal Trade in Ghanaian Urban Centres: Case Studies in Accra and Kumasi. Unpublished B.Sc. *Dissertation,* University of Science and Technology, Kumasi.

Jenik, J. and Hall, J.B. 1976. *Plant pommunities of the Accra Plains*. Folia Geobot Phytotax Praha, 162-212.

Johnson, H. 1992) Rural livelihoods: action from below. In Berstein, H., Crow, B. and Johnson H. 1992. *Rural Livelihoods: Crises and Responses,* 274-306. Oxford University Press, Oxford.

Johnson, N.L. and Ruttan V.W. 1994. Why are Farms So Small? *World Development* 22(5): 691-706.

Johnson, S.D. 1992. Plant animal relationships. In Cowling, R.M. 1992. *The Ecology of Fynbos: Nutrients, Fire and Diversity*, 175-205. Oxford University Press, Cape Town.

Jones, R.G.B. 1975. Report on a soil conservation consultancy to study soil erosion problems on the Jos Plateau. *Land Resources Report No. 6.* Ministry for Overseas Development, U.K.

Judson, S. 1981. What is Happening to Our Continents. In Skinner, B.J. (Ed.) *Use and Misuse of Earth's Surface,* 12 - 139. Los Altos, California.

Jurion, F. and Henry, J. 1969. *Can Primitive Farming be Modernised ?* Institut National pour l Etude Agronomique du Congo, Brussels.

Kalpage, F.S.C. 1974. *Tropical Soils Classification, Fertilty and Management.* Macmillan Press Ltd., London.

Kandor, D. 1993. *Waste management: how is waste management department facing the challenge?* The Peoples Daily Graphic. 10 February, 7.

Kang, B.T. 1986. Cropping systems and soil fertility management in the humid tropics with special reference to West Africa. *Developments in Plant and Soil Sciences* 24: 83 - 94.

Kang, B.T., Gichuru, M., Hulugalle, N. and Swift M.J. 1991. Soil Constraints for Sustainable Upland Crop Production in Humid and Sub - Humid West Africa. In Soil Constraints on Sustainable Plant production in the Tropics: Proceedings of the 24th International Symposium in Tropical Agriculture Research. Kyoto, August 14 - 16, 1990. *Tropical Agriculture Research Series* 24: 101 -102.

Karanth, G.K. 1995. *Change and Continuity in Agrarian Relations.* Concept Publishing Company, Dehli.

Karar, E. 2004. *Institutional change in river basin management and its implications for livelihood and ecosystem security in South Africa.* SIWI/CA Seminar. Stockholm, August 21.

Kassanga, R.K. 1992. *Agricultural Land Administration and Social Differentiation: A Case Study of the Tono, Vea and Fumbisi Belt of North-Eastern Ghana.* Social Science Research Council, New York.

Kates, R. 1987. The Human environment: the road not taken, the road still beckoning. *Annals of the Association of American Geographers* 77(4): 525-534.

Kaufman, Y.J., Setzer, A., Justice, C., Tucker, C.J., Pereira, M.C., and Fung, I. 1990. Remote sensing of biomass burning in the tropics. In Goldhammer, J.G. (Ed.) *Fire on the Tropical Biota: Ecosystem Processes and Global Challenges* 371-399, Springer Verlag, Berlin

Kaul, R. 1985. Tillage needs of small farmers in a developing country. In *Proceedings, International Conference on Soil Dynamics and Commemoration of the 50th Aniversary of the National Tillage Machinery Laboratory.* 559 - 558. Auburn University Press, Auburn.

Kay, G.R. 1972. *The Political Economy of Colonialism in Ghana: a Collection of Documents and Statistics 1900-1960.* Cambridge University Press, London,.

Keay, R.W.J. 1959. *Vegetation Map of Africa south of the Tropic of Cancer: Explanatory Notes*. Oxford University Press, London.

Khasa, P.D. and Dancik, B.P. 1997. *Sustaining tropical forest biodiversity. Journal of Sustainable Forestry* 5(1-2): 217-234.

Killick, T. 1978. *Development Economics in Action: A Study of Economic Policies in Ghana.* Heinemann, London.

Kodva, V. A. 1977. *Soil loss: an overview. Agro-Ecosystems* 3: 205-224.

Kodva, V.A. 1981. *Loss of productive land due to salinisation. Ambio* 12: 91 - 93.

Kollman, J. 2002. *Game theoretical analysis of the irrigation system in the Upper East Region of Ghana.* University of Bonn, Bonn.

Koomson, D. 1988. *Development and environment.* The People's Daily Graphic. 31 August, 5.

Koranteng, K.A. 1990. *Ghana Canoe Frame Survey 1989.* Fisheries Department Research and Utilization Branch, Tema.

Kowal, J.M. and Kassam A. H. 1977. Energy load and instantaneous intensity of rainstorms at Samaru, Northern Nigeria. In Greenland, D.J. and Lal, R. (Ed.) *Soil Conservation and Management in the Humid Tropics,* 57 - 77. John Wiley and Sons, Chichester.

Kumar, H. 1969. *Statistics of Large Scale Specialised, Institutional and Co-operative Farming.* Ministry of Agriculture, Accra.

Kwamena Poh, M.A. 1973. *Government and Politics in the Akwapim State 1730 -1850.* North Western University Press/Longman, London.

La Anyane, S. 1956. An historical account of agricultural development in the Gold Coast before 1900. *New Gold Coast Farmer* 1: 28-65.

Laing, E. 1994. *Ghana Environmental Technical Background Papers by the Working Groups.* Government of Ghana, Accra.

Lal, R. 1976. *Soil erosion Investigations on an Alfisol in Southern Nigeria.* I.I.T.A. Monograph No.1., Ibadan.

Lal, R. 1985. Soil erosion and its relation to productivity in tropical soils. In El-Swaify, S.A. Moldenhauer P. and Lo, A. (Eds) *Soil Erosion and Conservation, 237 - 247.* Soil Conservation Society of America, Ankeny.

Lal, R. 1988. Soil Erosion Research Methods. *Soil Conservation Society,* Ankeny.

Lal, R. 1990a. Soil erosion and land degradation: the global crisis. *Advances in Soil Science* 11: 129 - 172.

Lal, R. 1990b. *Soil Erosion in the Tropics: Principles and Management.* McGraw Hill Inc. New York.

Lal, R. 1993. Soil erosion and conservation in West Africa. In Pimemtal, D. (Ed.) *World Soil Erosion and Conservation,* 7 - 25. Cambridge University Press, Cambridge.

Lane, D.A. 1962. The Forest Vegetation. In Wills, J.B. 1962. *Agriculture and Land Use in Ghana.* Oxford University Press, London.

Larsen, R.E. 1968. Is There a future for livestock production in Ghana? *The Ghana Farmer* XII(2): 52-54.

Lassailly-Jacob, V. and McDowell C. 1996. Land-based strategies in dam-related resettlement programmes in Africa. In McDowell, C. 1996. *Understanding Impoverishment: the Consequences of Development Induced Displacement,* 187-199. Berghahm Books, Providence.

Lawry, S. 1990. *Tenure Policy and Natural Resource Management in Sahelian West Africa.* Land Tenure Center, Madison.

Lawson, G. 1986. *Plant Ecology in West Africa: Systems and Processes.* John Wiley and Sons, Chichester.

Lawson, R.M. 1968. The transition of Ghana's fishing from a primitive to a mechanised industry. *Transactions of the Historical Society of Ghana IX:* 90 - 96.

le Maitre, D.C. and Midgley, J.J. 1992. Plant peproductive ecology. In Cowling, R.M. 1992. *The Ecology of Fynbos: Nutrients, Fire and Diversity,* 135-174. Oxford University Press, Cape Town.

Leach, M. and Fairhead, J. 2000. Challenging neo-Malthusian deforestation analysis in West Africa's dynamic forest landscapes. *Population and Development Review* 26: 11–43.

Leach, M., Mearns, R. and Scoones, I. 1997. *Environmental Entitlements: a Framework for Understanding the Institutional Dynamics of Environmental Change.* University of Sussex, Brighton.

Ledant, J.P. 1985. The reduction of biomass in West Africa - Part two - Assessment. *Annales - de Gemblox.* 91 (2): 111 - 123.

Ledger, D.C. 1964. Some hydrological characteristics of West African rivers *Transactions of the Institute of British Geographers* 35: 73 - 90.

Leff, E. 1995. *Green production.* Guilford Press, New York.

Lelong, F., Roose, E., Aubert, G., Fauck, R. and Pedro, G. 1984. Present day geodynamics in different West African soils under cultivation or natural vegetation. *Catena* 11(4): 343 - 376.

Lemly, D. A, Kingsford, R. T., Thompson, J. R. 2000. Irrigated agriculture and wildlife conservation: conflict on a global scale. *Environmental Management* 25: 485-512.

Lieberman, D.D. 1979. *Dynamics of Forest and Thicket Vegetation on the Accra Plains, Ghana.* Unpublished Ph.D Thesis, University of Ghana.

Linder, H.P., Meadows, M.E. and Cowling, R.M. 1992. History of the Cape flora. In Cowling, R.M. 1992. *The Ecology of Fynbos: Nutrients, Fire and Diversity,* 113-134. Oxford University Press, Cape Town.

Lipton, M. 1982. Why poor people stay poor. In Harriss, J. (Ed.) *Rural Development Theories of Peasant Economy and Agrian Change*, 66 - 81. Hutchinson, London.

Little, P. 1992. *The Elusive Granary: Herder, Farmer and State in Northern Kenya.* Cambridge University Press, Cambridge.

Long, N. 1992. From paradigm lost to paradigm regained? The case for an actor-oriented sociology of development. In Long, N. and Long, A. (Eds.) Battlefields of Knowledge: *The Interlocking of Theory and Practice in Social Research and Development,* 16-43. Routledge, London.

Long, N. and Van der Ploeg, J.D. 1994. Heterogeneity, Actor and Structure: Towards a Reconstution of the Concept of Structure. In Booth, D. (Ed.) *Rethinking Social Development: Theory, Research and Practice*, 62-89. Longman, Harlow.

Longman, K.A. and Jenik, J. 1992. Forest-Savanna Boundaries: General Considerations. In Furley, P.A., Proctor, J. and Ratter, J.A. 1992. *Nature and Dynamics of Forest-Savanna Boundaries,* 3-20. Chapman and Hall, London.

Lowdermilk, W.C. 1953. *Conquest of the Land Through Seven Thousand Years.* SCS Agriculture Information Bulletin, 99 Washington.

Lowe, M.S. and Bowlby, S.R. 1992. Population and the environment. In *Mannion, A.M. and Bowlby, S.R. (Eds.), 117- 130.* John Wiley and Sons, Chichester.

Loxton, R.F. Hunting and Associates 1974. *The Natural Resources of Damaraland.* Unpublished Report undertaken for the Department of Bantu Administration and Development, Pretoria.

Maass, J.M. 1995. Conversion of dry tropical forest to pasture and agriculture. In Bullock, S.H., Mooney, H.A. and Medina, E. (Eds) *Seasonally Dry Tropical Forests, 399-422.* Cambridge University Press, Cambridge.

Mabberly, D.J. 1992. *Tropical Rainforest Ecology.* Blackie, Glasgow.

Macdonald, G. 1898. *The Gold Coast Past and Present: A Short Description of the Country and its People.* Longman, Green and Company, London.

Malingreau, J.P. 1990. The contribution of remote sensing to the global Monitoring of fires in tropical and sub-tropical ecosystems. In Goldhammer, J.G. (Ed.) *Fire in the Tropical Biota, Ecosystem Processes and Global Challenges,* 337-369. Springer Verlag, Berlin.

Malthus, T. 1798. An Essay on The Principle of Population and a Summary View of the Principle of Population. Cited in Malthus, T. and James, P. 1990. *An Essay on The Principle of Population.* Cambrige University Press, Cambridge.

Mankure, E. 1970. *Economic Organisation the Poultry Industry of Ghana* Unpublished M.Sc. Dissertation, University of Ghana, Accra.

Manshard, W. 1974. *Tropical Agriculture: An Geographical Introduction and Appraisal.* Longman, London.

Marchal, J.K. 1983. *Yatenga Nord - Haute - Volta. La dynamique d' un espace rural soudano - sahelien.* 167, Trav Doc Orstrom.

Marchand, M.H. and Parpart, J.l. (Eds) 1995. *Feminism/Post Modern/Development.* Routledge, London.

Margelef, R. 1968. *Perspectives in Ecological Theory.* University of Chicago Press, Chicago.

Marius, C. and Lucas, J. 1991. Holocene mangrove swamps of West Africa: sedimentology and soils. *Journal of African Earth Sciences,* 12 (1 - 2), 41 - 54.

Marks, I. 1996. Women and energy resources management: a UNIFEM perspective. *Natural Resources Forum* 20 (2):145-152).

Marshall, G. 1994. *The Concise Oxford Dictionary of Sociology.* Oxford University Press, Oxford.

Martinez-Alier, J. 1992. *Ecologia y Pobreza.* Centre Cultural Bancaixa, Barcelona.

Mason, J.J. and Danso, E.Y. 1995. Participatory Rural Appraisal for People and Parks: the Case of Mole National Park, Ghana. *PLA Notes* 22: 76-79. IIED, London.

Matlon, P.J. 1987. The West African semi arid tropics. In Mellor, J.W. and Blackie, M.J. (Eds.) *Accelerating Food Production in Subsaharan Africa,* 39 - 77. John Hopkins Press, Baltimore.

Mayoux, L. 1995. Beyond naivety: women, gender inequality and participatory development. *Development and Change* 26 (2): 235-258.

McCarthy, J.F. 2002. First world political ecology: lessons from the wise use movement. *Environment and Planning A* 34: 1281-302.

McCarthy, J.F. 2005. Guest editorial. *Environment and Planning* A 37: 953-958.

McCarthy, J.F. 2006. *The Fourth Circle. A Political Ecology of Sumatra's Rainforest Frontier.* Stanford University Press; Palo Alto.

McNeely, J.A. 1993. *Lessons from the Past:* Forests and Biodiversity Mimeo Gland, IUCN.

Meadows, D.H., Meadows, D.L. Randers, J. and Behrens III, W.W. 1972. *The Limits to Growth: A Report to the Club of Rome's Project on the Predicament of Mankind.* Potomac Associates, New York.

Medina, E. and Huber, O. 1994. The Role of biodiversity in the functioning of savanna ecosystems. In Solbrig, O.T., Van Emden, H.M. and Oordt, P.G.W.J. (Eds.) *Biodiversity and Global Change*, 141-160. CAB International, Wallingford.

Mellor, M. 1997. Gender and the environment. In Redclift, M. and Woodgate, G. (eds.) *The International Handbook of Environmental Sociology,* 367-385. Edward Elgar, Cheltenham.

Menaut, J.C. 1983. The vegetation of African savannas. In Boulrliere, F. (Eds.). *Tropical Savannas.* Elsevier Scientific Publishing Company Amsterdam.

Menaut, J.C., Gignoux, J., Prado, C. and Clobert J. 1990. Tree community dynamics in a humid savanna of the Cote de Ivoire: modelling the effects of fire and competition with grass and neighbours. *Journal of Biogeography* 17: 471-481.

Mends, H. 1955. Agricultural policy in Ghana. *The Ghana Farmer* XVII(2): 47-48.

Merchant, C. 2005. *Radical Ecology: in search of a Liveable World.* Routledge, London.

Mertens, W. 1994. *Population and Deforestation in Humid Tropics.* International Union for the Scientific Study of Population, Ghent.

Middleton, N.J. 1985. Effect of drought on dust production in the Sahel. *Nature* 316 (6027): 431 - 434.

Miedema, R., Jongmans, A.G., and Brinkman, R. 1987. The Micromorphology of a typical catena from Sierra Leone, West Africa. *Soil Micromorphology 137 - 144, Association Francaise pour l Etude du Sol,* Versailles.

Migot-Adholla, S.E. 1994. Land, secuirity of tenure, and productivity in Ghana. In Bruce, J.W., Benneh, G., Place, F., Atsu, S. (Eds.) *Searching for Land Tenure Security in Africa,* 97-118 Kendall/Hunt Publishing Company, Dubuque.

Mikkelsen, J.H. and Langohr, R. 2004. Indigenous knowledge about soils and sustainable crop production, a case study from the Guinea woodland savannah (Northern Region, Ghana). *Danish Journal of Geography* 104(2): 12 – 26.

Milas, S. 1984. Population crisis and desertification in the Sudano - Sahelian Region. *Environmental Conservation.* 11 (2), 167 -169.

Miller, R.W. and Donahue, R.L. 1990. *Soils: An Introduction to Soil and Plant Growth.* Prentice Hall International Inc., London.

Miller, G.T. Jr and Hackett, D.2011. *Living in the Environment.* 2nd ed. Nelson Education, Toronto.

Milleville, P. 1980. *Etude d'un systeme de production agro pastoral sahelien de Haute - Volta.* I. Le systeme de culture. Orstrom Ouagadougou.

Millington, A.C. 1992. *Soil Erosion and Conservation. In Mannion, A.M. and Bowlby,* S.R. (Eds.), 227 - 243, John wiley and Sons, Chichester.

Millington, A.C., Mutiso, S.K., and Kirkby, J. and O'Keefe, P. 1989. African soil erosion - nature undone and the limitation of technology. *Land Degradation and Rehabilitation* 1, 279 - 290.

Millington, A.C., Styles, P.J. and Critchley, R.W. 1992. *Mapping Forests and Savannas in Sub-Saharan Africa From Advanced Very High Resolution (AVHRR) Imagery.* In Furley, P.A., Proctor, J. and Ratter, J.A. 1992. Nature and Dynamics of Forest-Savanna Boundaries, 37-62. Chapman and Hall, London.

Mill, J.S.1867. Book IV quoted in Common, M. 1988. *Environmental and Resource Economics: An Introduction,* 6. Longman London.

Milner, C. and Douglas, M.G. 1989. *Problems of Land Degradation in Commonwealth Africa: Study on the Scope for Common Action. Food Production and Rural Development Division, Commonwealth Secretariat,* London.

Mitchell, C.W. 1984. *Soils. Key Environments: Sahara Desert.* 48 - 55, Pergamon Press, Oxford.

Mohan, G. 1996. Neoliberalism and decentralisation development planning in Ghana. *Third World Planning Review* 18(4): 433-454.

Monu, E.D. 1997. Farmer participation in research: implications for agricultural development. *Journal of Social Development in Africa* 12 (1):53-66.

Moock, J.I. and Rhoades, R.E. (eds) 1992. *Diversity, Farmer Knowledge, and Sustainability.* Cornell University, Ithaca.

Moore, D.S. 1996. Marxism, Culture and Political: Environmental Struggles in Zimbabwe's Eastern Highlands. In Peet, R. and Watts, M.(Eds.) 1996. *Liberation Ecologies: Environment, Development, Social Movements.* Routledge, London.

Moran, M.H. 1990. *Civilized Women: Gender and Prestige in Southeastern Liberia.* Cornell University Press, Ithaca.

Mortimore, M. and Tiffen, M. 1995. Population and Environment in Time Perspective: the Machakos Story. In Binns, T. 1995. (Ed.) *People and Environment in Africa.* John Wiley and Sons, Chichester.

Motha, R.P., Leduc, S.K., Steyaert, L.T., Sakamota, C. M. and Strommen, N.D. 1990. *Precipitation patterns in West Africa.* Monthly Weather Review, 108: 1567 - 1578.

Mueller-Dombois, D. and Goldhammer, J.G. 1990. Fire in tropical ecosystems and global environmental change: an introduction. In Goldhammer, J.G. 1990. *Fire in the Tropical Biota: Ecosystems and Global Challenges,* 1-10. Spriger Verlag, Berlin.

Munasinghe, M. 1993. The economist's approach to sustainable development. *Finance and Development* 30(4): 6 - 19.

Murdoch, J. 1998. *The spaces of actor network theory. Geoforum* 294: 357 – 374.

Musa, I., 1988. *Land Degradation in the Bawku East Area.* Unpublished B.A. Dissertation, University of Ghana, Accra.

Muylwijk, J. 1996. *Bibliography on Gender and Agricultural Engineering for Small Farms in Developing Countries.* AGSE Occasional Paper WAO/FAO, Rome.

Muylwijk, J. and Smetsers, M. 1996. Gender and Agricultural Engineering: an Overview of Current Theory and Praxis, *Focusing on Sub-Saharan Africa with Case Studies from Zimbabwe and Senegal.* AGSE Occasional Paper WAO/FAO, Rome.

Nabila, J.S. 1986. *Rural Migration and its Implications for Rural Development in Ghana.* In Brown, C.K. (Ed.) Rural Development in Ghana, 75 - 89, Ghana Universities Press, Accra.

Natural Resources Institute 1996. Women and Food Processing in Sub-Saharan Africa. *Proceedings of a Workshop Arranged Jointly by Women in Agricultural Development (WIAD),* Ministry of Agriculture, Ghana and the European Consortium for Agricultural Research in the Tropics (ECART) held in Accra, Ghana, 6-9 December 1994. Natural Resources Institute, Chatam.

Neumann R P 1998. *Imposing wilderness: struggles over livelihood and nature preservation in Africa.* University of California Press, Berkeley.

Ngere, I.O. 1971. Adapting livestock improvement programs to the farmers needs. *Legon Agricultural News* 4(3): 29 - 32.

Ngwa, N.E. 1995. The role of women in environmental management: an overview of the rural Camerounian situation. *GeoJournal* 35 (4):515-520.

Nicholson, S.E. 1980. The nature of rainfall fluctuations in subtropical West Africa. *Monthly Weather Review,* 108, 413 - 487.

Nicolis, G. 1994. *Dynamical systems, biological complexity and global change.* In Solbrig, O.T., Van Emden, H.M. and Oordt, P.G.W.J. (Eds.) Biodiversity and Global Change, 21-31. CAB International, Wallingford.

Nielsen, M.S. 1965. *An Introduction to the Flowering Plants of West Africa.* University of London Press, London.

Nelson, V. and Stathers, T. 2009. Resilience, power, culture, and climate: A case study from semi-arid Tanzania, and new research directions. *Gender & Development.* 17(1), 81-94.

Nietschamann, B. 1991. Conservation by Self-Determination *National Geographic Research and Exploration* 7 (3): 372-373.

Niewolt, S. 1977. *Tropical Climatology: An Introduction to the Climates of the Low Latitudes.* John Wiley and Sons London.

Nisiah Gyabaah, K. 1992. *Environmental Degradation and the Threat of Drought and Desertification in the Upper West Region of Ghana: a Search for Sustainable Development Solutions.* Unpublished Ph.D Thesis University of London.

Norgaard, R.B. 1994. *Development Betrayed: The End of Progress and a Coevolutionary Revisioning of the Future.* Routledge, London.

Norgaard, R.B. 1984. *Economic Development and Cultural Change.* Routledge, London.

Norman, M.J.T., Pearson, C.J., and Searle, P.G.E. 1984. *The Ecology of Tropical Food Crops.* Cambridge University Press, Cambridge.

Norman, W.R. 1997. Indigenous Community Managed Irrigation in Sahelian West Africa. *Agriculture, Ecosystems and Environment* 61 (2-3): 83-95.

Nortcliff, S. 1988. Soil formation and characteristics of soil profiles. In Wild, A. (Ed.) *Russell's Soil Conditions and Plant Growth,* 168 - 212. Longman, Harlow.

Norton, A. 1987. *The Socio Economic Background of Community Forestry in the Northern Region of Ghana.* Overseas Development Administration, London.

Ntimoa-Baidu, B. 1994. *Indigenous Versus Traditional Biodiversity Conservation Strategies: The Case of Protected Area Systems in Ghana.* Biodiversity Support Program, USAID, Washington.

Nye, P.H., and Greenland, D.J. 1960. *The Soil Under Shifting Cultivation, Commonwealth Bureau of Soils,* Harpenden.

Nyerges, A.E. 1987. The Development Potential of the Guinea Savanna: Social and Ecological Constraints in the West African 'Middlebelt'. In Little, P.D., Horowitz, M.M. and Nyerges, A.E. (Eds.) *Land at Risk in the Third World: Local Level Perspectives* (Monographs in Development Anthropology), 316 - 336. Westview Press, Boulder.

O'Connor, A. 1991. *Poverty in Africa: A Geographical Approach.* Belhaven Press, London.

O'Connor, M. 1994. *Is Capitalism Sustainable?* Guildford, New York.

O'Connor, J. 1998: *Natural causes: essays in ecological Marxism.* Guilford Press, New York.

O'Neil, J. 1993. *Ecology, Policy and Polities: Human Well Being and the Natural World.* Routledge, London.

Obosu Mensah, K. 1996. *Ghana's Volta Resettlement Scheme: the Long-Term Consequences of Post-Colonial State Planning.* International Scholars Publications, Bethsada.

Ocansey, J.M. 1985. *A Survey of Wood Energy Utilisation within a Rural Community Unpublished Diploma Dissertation,* University of Science and Technology, Kumasi.

Odotei, I. 1991. *Introduction of New Technology in the Artisanal Marine Fishing Industry in Ghana I: A Historical Overview.* II The Role of Women. Senter for Milj og Utvikling, Trondhiem.

Oduro-Afriyie, K. 1996. *Rainfall erosivity map for Ghana.* Geoderma 74(1-2):161-166.

Ofomata, G.E.K. 1964. Soil erosion in the Enugu Region of Nigeria. In *African Soils* 9: 321 - 348.

Ofori, M. 1970. An integrated approach to rural development. *Legon Agricultural News* 4(4): 8-15.

Ohene-Konadu, K. 1996. Why rural people in Ghana have achieved so little? *Journal of Rural Development and Adminstration* 28 (1):63-77.

Okigbo, B.N. 1977. Farming systems and soil erosion in West Africa. In *Soil Conservation and Management in the Humid Tropics,* 151 - 163. John Wiley and Sons, Chichester.

Okyere, W., Doss, C.R. and Olson C. 1991. The Effects of Domestic Policies on Exportable Primary Commodities: the Case of Ghana and Cocoa. *Issues in African Rural Development 21-47 Winrock International Institute for Agricultural Development*, Arlington.

Olayide, S.O. and Falusi, A.O. 1977. Economics of soil conservation and erosion and control practices in Nigeria. In *Soil Conservation Management in the Humid Tropics,* 117 - 125. John Wiley and Sons, Chichester.

Oldman, L.R. 1994. The global extent of soil erosion degradation. In Greenland, D.J. and Szabolcs,S.J. (Eds.) *Soil Resilience and Sustainable Land Use,* 99 - 118. C.A.B. International, Oxford.

Oliver, R. 1995. Contraceptive use in Ghana: the role of service availability, quality and price. *Living Standards Measurement Study Working Paper* No. 111, World Bank, Washington DC.

Olson, G.W. 1981. Archaeology: lessons on future soil use. *Journal of Soil Water Conservation.* 36 (5): 261 - 264.

Ophuls, W. 1977. *Ecology and the Politics of Scarcity.* Freeman, San Francisco.

Opoku Afriyie, Y. 1986. The Nature of infrastructure for rural development. In Brown, C.K. (Ed.) *Rural Development in Ghana,* 47-63. Ghana University Press, Accra.

Opoku-Ankromah, Y. 2000. *Impact, Vulnerability and Adaptation Assessment in Water Resources.* Accra, Environmental Protection Council.

Oppong, E.N.W. 1970. *The livestock industry in the Accra Plains.* Legon Agricultural News 3(2): 1 - 14.

O'Riordan, T. 1981. *Environmentalism.* Pion, London.

O'Riordan, T. 1989. The challenge for environmentalism. In Peet, R. and Thrift, N. (Eds). *New Models In Geography.* Unwin Hyman, London.

Ortner, S. 1974. Is female to male as nature is to culture? In Rosaldo, M. and Lamphere, L.(Eds.) 1974. *Woman, Culture and Society,* 67-87. Stanford University Press, Stanford.

Osei, W.Y. 1993. Woodfuel and deforestation - answers for a sustainable environment. *Journal of Environmental Management* 37 (1): 51-62.

Osei, W.Y. 1996a. Socio-economic dynamics of forest loss. *Rural Agroecosystems in Environmentalist* 16 (3): 231-239.

Osei, W.Y. 1996b. Rural energy technology: issues and options for sustainable development in Ghana. *Geoforum* 27 (1):63-74.

Otoo, J. 1993. *Waste disposal in Accra: a challenge to the Accra Metropolitan Authority.* The Peoples Daily Graphic 5 February, 5.

Owusu-Ansah, David. 1995. Urban-Rural Disparities. In Berry, L. B. (Ed.) *A Country Study: Ghana. United States Government t Printing Office,* Washington, DC.

Owusu - Ansah, K.A. 1986. Managing rural socio-economic development in Ghana. In Brown, C.K. (Ed.) *Rural Development in Ghana,* 237 - 261. Ghana Universities Press, Accra.

Pacey, A. and Payne, P.(Eds.) 1985. *Agricultural Development and Nutrition.* Hutchinson, London.

Packman, J.R., Harding, D.J.L., Hilton, G.M. and Stultard, R.A. 1992. *Functional Ecology of Woodlands and Forests.* Chapman and Hall, London.

Painter, M.and Durham, W. 1995. (Eds.) *The Social Causes of Environmental Destruction in Latin America.* University of Michigan Press, Ann Arbor.

Palo, M., Yirdaw, E. and Mery, G. 1996. Deforestation and Sustainable Forestry Challenge in Ghana. In Palo, M. 1996. *Sustainability Forestry Challenges for Developing Countries,* 343-357. Kluwer Academic Publishers, Dordrecht.

Papanek, H. 1990. To each less than she needs, from each more than she can do: allocations, entitlements, value. In Tinker, I. (Ed.) *Persistent Inequalities, Women and World Development,* 162 - 181. Oxford University Press, New York

Parker, R.L., Riopelle, R. and Steel, W.F. 1995. *Small Enterprises Adjusting to Liberalisation in Five African Countries.* World Bank, Washington DC.

Parren, M.P.E. and Graaf, N.R. 1995. *The Quest for Natural Forest Management in Ghana,* Coted'Ivoire and Liberia. Stichting Tropenbos, Wageningen.

Paulino, L.A. 1986. *Food in the Third World.* I.F.P.R.I. Resources Report 52, Washington.

Paulson, S., Gezon, L.L. and Watts, M. 2003. Locating the political in political ecology: An Introduction. *Human Organization* 62(3): 205-217.

Pearce, R. 1992. Ghana. In Duncan, A. and Howell, J. (Eds.) *Structural Adjustment and the African Farmer*, 14 - 40, Overseas Development Institute, London.

Pearce, D., Markandya, A. and Barbier, E. 1989. *Blue Print for a Green Economy.* Earthscan, London.

Pearce, D., Markandya, A.and Barbier, E. 1990. *Sustainable Development: Economics and Environment in the Third World.* Edward Elgar Publishing Ltd., Aldershot.

Pearce, D. and Turner, R.K. 1990. *Economics of Natural Resources and the Environment.* Harvester-Wheatsheaf, New York.

Peet, R. and Watts, M.J. 1993. Introduction: development theory and environment in an age of market triumphalism. *Economic Geography* 68(3): 227-253.

Peet, R. and Watts, M. 1996. *Liberation Ecologies: Environment, Development, Social Movements.* Routledge, London.

Pelisser, P. 1966. *Les paysans au Senegal. Les Civilisations agraires du Cayor a la Casamance.* Fabregue, Saint-Yrieix.

Peluso, N. 1992. *Rich Forests, Poor People: Resource Control and Resistance in Java.* University of California Press, Berkeley.

Pepper, D. 1996. *Modern Environmentalism: A Introduction.* Routledge, London.

Perry, R.J. 2003. *Five Key Concepts in Anthropological Thinking. Upper Saddle River,* Prentice Hall, New Jersey.

Peters, P. 1994. *Dividing the Commons: Politics, Policy and Culture in Botswana.* University Press of Virginia, Charlottesville.

Peters, R.H. 1991. *A Critique of Ecology.* Cambridge University Press, Cambridge.

Peterson, V.S., & Runyan, A.S. 2010. *Global Gender Issues in the New Millennium.* Westview Press, Boulder.

Pettersen, S. 1969. *Introduction To Meteorology.* (3rd ed). McGraw Hill, New York.

Philips, J.F.V. 1966. *The Development of Agriculture and Forestry in the Tropics.* (2nd ed.), Faber and Faber, London.

Pickett, S.T.A., Parker, V.T. and Fiedler, P.L. 1992. The New Paradigm in Ecology: Implications for Conservation Biology above the Species Level. I Fiedler, P.L. and Jain, K. 1992. *Conservation Biology: The Theory and Practice of Nature Conservation, Preservation and Management,* 66-87 Chapman and Hall, New York.

Pimentel, D., Allen, J., Beers, A., Guinand, L., Hawkins, A., Linder, R., McLaughlin, P., Meer, B., Musonda, D., Perdue, D., Poisson, S., Salazar, R., Siebert, S. and Stoner, K. 1993. Soil Erosion and Agricultural Productivity. In Pimentel, D. (Ed.) *World Soil Erosion and Conservation.*, 277 - 292. Cambridge University Press, Cambridge.

Pimm, S.L 1991. *Balance of Nature: Ecological Issues in the Conservation of Species and Communities.* University of Chicago Press, Chicago.

Platteau, J.P. 1997. Reforming land rights in sub-Saharan Africa: an issue of efficiency and equity. *Journal fur Entwicklungspolitik* (1):-98.

Plumwood, V. 992. *Beyond the Dualistic Assumptions of Women, Men and Nature Ecologist* 22(1): 8-13.

Pogucki, R.J.H. 1955. *Gold Coast Land Tenure 3: Land Tenure in Ga Customary Law.* Government Printer, Accra.

Pomeroy, J.Y. 2012. *Analyzing Municipal Annexations: Case Studies of Frederick and Caroline Counties of Maryland 1990 - 2010.* Unpublished PhD Thesis, University of Maryland, College Park.

Pound, B., Snapp, S., McDougall, C. and Braun, A. (Eds.) 2003. *Managing Natural Resources for Sustainable Livelihoods: Uniting Science and Participation.* James & James/Earthscan, London.

Powell, J.M, Pearson, R.A. and Hiernaux, P.H. 2004. Crop–livestock interactions in the West African drylands. *Agronomy Journal* 96: 469-.)483.

Powell, J.W. 1986. *The Role of a Technological University in Integrated Rural Development.* In Brown, C.K. (Ed.) Rural Development in Ghana., 181 - 187. Ghana Universities Press, Accra.

Pretty, J.N. 1995. *Participatory Learning for Sustainable Agriculture World Development* 23(8): 1247-1263.

Pretty, J.N. and Shah, P. 1997. Making soil and water conservation sustainable: from coercion and control to partnerships and participation. *Land Degradation and Development* 8 (1):39-58.

Quansah, C. 1981. The effect of soil type on slope, rain intensity and their interactions on splash detachment and transplant and transport. *Journal of Soil Science* 32: 215 - 224.

Rao, B. 1991. Dominant Constructions of Women and Nature in Social Science Literature, Capitalism, *Nature,* Socialism Pamphlet 2. Guilford Publications, New York.

Rawls, J. 1972. *A Theory of Justice Oxford.* University Press, Oxford.

Raymond, G. and Tourte, R 1985. *Note de synthese sur la mise en place du project federateur de recherche-develoment au Togo.* Rapport de mission du 1 au 10 octobre 1985. IRCT-DSA, Montpellier.

Raynaut, G. 1980. Rescherhes multi disciplinaires sur la region de Maradi. Rapport de synthese. Univ Bordeaux ll - DGRST. Cited in *Fertility of Soils: A Future for Farming in the West African Savanna.* (C.J.M.G. Pieri) 1992, 88, Springer Verlag, Berlin.

Razavi, S. and Miller, C. 1995. *From WID to GAD: Conceptual Shifts in the Women and Development Discourse.* United Nations Research Institute for Social Developent, New York.

Reardon, T. and Vosti S.A. 1995. Links between rural poverty and the environment in developing ccountries: asset categories and investment poverty. *World Development* 23(9): 1495-1506.

Redclift, M. 1987. *Sustainable Development: Exploring the Contradictions* Methuen, London.

Redclift, M. 1993. Sustainable development: needs, values, rights. *Environmental Ethics* 2 (1):3-20.

Redclift, M. and Goodman, D. 1991. Introduction. In Goodman, D. and Redclift, M. (Eds.) *Environment and Development in Latin America: The Politics of Sustainability, 1 - 23.* Manchester University Press, Manchester.

Redclift, M. and Woodgate, G.W. 1993. Concepts of the Environment in the Social Sciences. In *Wye College External Programme* - M.Sc. in Enviromental Management, Course File, Units 6 and 10.

Reenberg, A. and Paarup Laursen, B. 1997. Determinants for Land use strategies in a Sahelian agroecosystem: anthropological and ecological geographical aspects of natural resource management. *Agricultural Systems* 53 (2-3) 209-229.

Rees, C. 1993. The ecologist's approach to sustainable development. *Finance and Development.* December, 14 to 15.

Reijntjes, C., Haverkort, B. and Waters - Bayer, A. 1992. *Farming for Future: An Introduction to Low - External Input and Sustainable Agriculture.* Macmillan, ILEA, Leusden.

Reppetto, R., and Gills, M. 1988. *Public Policies and the Misuse of Forest Resources. Cambridge* University Press, Cambridge.

Republic of Ghana 1989. 1984 Population Census of Ghana: Special Report on Localities by Local Authorities. *Greater Accra Region.* Statistical Service, Accra.

Richards, P. 1985. *Indigenous Agricultural Revolution.* Hutchinson, London.

Richards, P. and Deimer, G. 1996. Agrarian Technologies as socio-technical hybrids. *Food crop improvement and management of land and water in sub-Saharan Africa.* Bulletin - APAD 11: 17-36.

Riddell, R. 1981. *Ecodevelopment.* Gower Publishing Co. Ltd., Westmeal.

Riehl, H. 1954. *Tropical Meteorology.* McGraw Hill, New York.

Riley, J. and Alexander, C.J. 1997. *Statistical Literature for Participatory on-Farm Research in Experimental Agriculture* 33 (1):73-82.

Rimmer, D. 1992. *Staying Poor: Ghana's Political Economy* 1950-1990. Pergamon Press, Oxford.

Ritzer, G. 2008. *Modern Sociological Theory.* McGraw-Hill, Boston.

Robbins, P. 2004. *Political Ecology: A Critical Introduction.* Blackwell Publishing, Malden.

Rocheleau, D. 1991. Gender, ecology, and the science of survival. *Agriculture and Human Values* 8: 156-165.

Rocheleau, D. 1995. Gender and a feminist political ecology perspective. *Institute for Development Studies* 26(1): 9-16.

Rocheleau, D., Thomas-Slayter, B. and Wangari, E. 1996. *Feminist Political Ecology.* Routledge, London.

Roose, E.J. 1983. Runoff and erosion before and after cleaning in relation to cropping in West Africa. *Pedologie* 30 (4): 327 - 339.

Roose, E.J. 1984. Incidence of land clearing on the deterioration of tropical soils. *Machinisme - Agricole - Tropical* 87: 24 - 36.

Roose, E.J. 1986. Runoff and erosion before and clearing depending on the type of crop in western Africa. In *Land Clearing and Development in Tropics, 317* - 329. A.A. Balkenna, Rotterdam.

Roose, E.J. 1989. Soil and water conservation in the Sudano - Sahelian Zone. In Isherwood, K.F. (Ed.) *Fertiliser Use and Crop Production*, 157 - 165. International Fertiliser Industry Association, New York.

Rose Innes, R. 1977. *A Manual of Ghana Grasses.* Ministry of Overseas Development, London.

Rukuni, M. 1995. Policy Options for Irrigated Food Production in Southern Africa. *Journal of Applied Science in Southern Africa* (JASSA)1(1): 89-95.

Runnalls, D. 1986. *Environment and Development: a Critical Stocktaking.* North South Institute, Montreal.

Russell, S. 1999. *Landuse Changes and Rural Livelihoods: A Case Study in Nyambwani village,* Malawi. Goteborg University, Goteborg.

Ryszkowski, L. and Jankowiak, J. 2002. Development of agriculture and its impacts on landscape functions. In Ryszkowski, L. ed. Landscape *Ecology in Agroecosystems Management, 9-27.* CRC Press. Boca Raton.

Sachs, W. 1992. Environment. In Sachs, W. (Ed) *The Development Dictionary,* 26-37. Zed Books, London.

Sage, C. 1994. Population Consumption and Sustainable Development. In *Strategies for Sustainable Development: Local Agendas for the Southern Hemisphere* (eds. M. Redclift and C. Sage) 35 - 56, John Wiley and Sons, London.

Samanta, R.K. 1995. *Women in Agriculture: Perspective, Issues and Experiences.* M.D. Publications, New Dehli.

Sanford,W.W. and Isichei, A.O. 1986. Savanna. In Lawson, G. 1986. *Plant Ecology on West Africa Systems and Processes,* 119 -136. John Wiley and Sons, Chichester.

Sarfo-Mensah, P. 2002. Is 'Tumi' in the sacred grove really gone? Local interpretations of change in the landscape of the forest-savanna transition. Ghana. *IDS* 33(1): 47 -54.

Sarpong, D. B. and Asuming-Brempong, S. 2004. Responding to economic shocks in Ghana: the agricultural sector as a social safety net. *Electronic Journal of Agricultural and Development Economics* 1: 117–137.

Sarris, A.H. and Shams, H. 1991. *Ghana Under Structural Adjustment: the Impact on Agriculture and the Rural Poor.* New York University Press, New York.

Satti, A.M. 1995. *Country Environmental Profile: Ghana* African Development Bank Abidjan.

Sayer, J.A., Harcourt, C.S., and Mark-Collins, N. 1992. *The Conservation Atlas of Tropical Forests: Africa.* Macmillan, London.

Schiewer, H. 1995. Incorporation of village level needs into district planning in Ghana. Planning from below or continuation of planning for elites? *Africa Spectrum* 30(3): 293 314.

Schmidt-Kallert, E. 1995. African Reality and Foreign Concepts. *Development and Cooperation* 1: 23-24.

Scholes, R.J. and Walker, B.H. 1993. *An African Savanna: Synthesis of the Nylsvley Study.* Cambridge University Press, Cambridge.

Schreiber, G. and Hill, I. 1994. Promoting community based natural resource management - a pilot approach under the environmental resource management project in Ghana. *Agriculture and Rural Development* 1 (1): 24-27.

Schuurman F.J. 1993. B*eyond the Impasse: New Directions in Development Theory.* Zed Books, London.

Scoones, I. 1997. The Dynamics of soil fertility change: historical perspectives on environmental transformation from Zimbabwe. *The Geographical* Journal 63(2): 161-169.

Scoones, I. 1996. Hazards and opportunities: farming livelihoods in dryland Africa: lessons from Zimbabwe. Zed Books, *International Institute for Environment,* London.

Scoones, I. and Thompson, J. (Eds.) 1994. *Beyond Farmer First:Rural People's Knowledge, Agricultural Research and Extension Practice.* Intermediate Technology Publications, London.

Scoones, I., Toulmin, C. and Lane, C. 1993. Land Tenure for Pastoral Communities. In Young, M.D., and Solbrig, O.T. (Eds.) *The Worlds Savannas: Economic Driving Forces, Ecological Constraints and Policy Options for Sustainable Land Use, 49-66.* Parthenon Publishing Group, Paris.

Scoones, I. 1994. *Living With Uncertainty:New Directions for Pastoral Development in Africa.* IT Publishers, London.

Scott, J.C. 1985. *Weapons of the Weak: Everyday Forms of Peasants Resistance* Yale University Press, New Haven.

Scott, J.C. 1989. Everyday Forms of Peasant Resistance. In Colburn, F.D. (Ed.) *Everyday Forms of Peasant Resistance,* 1-23. M.E. Sharpe Inc., New York.

Scribner, S. 1995. *Policies Affecting Fertility and Contraceptive Use: An Assessment of Twelve Sub-Saharan Countries.* World Bank, Washington DC.

Seers, D. 1969. The meaning of development. *International Development Review* 11: 2–6.

Sen, A.K. 1981. *Poverty and Famines: An Essay in Entitlement and Deprivation.* Clarendon Press, Oxford.

Sen, S. and Grown, C. 1987. *Development, Crises and Alternative Visions: Third World Women's Perspectives.* Monthly Review Press, New York.

Serageldin, I. 1993. Making Development Sustainable. *Finance and Development,* December 6 to 10.

Sevilla-Guzman, E. and Woodgate, G. 1997. Sustainable rural development: from agriculture to agroecology. *The International Handbook of Environmental Sociology, 150-183.* Edward Elgar, Cheltenham

Shirazu, H.A. 1996. *Irrigation Development in Ghana Ghanaian Times* 8-10-1996, p.6.

Shiva, V. 1989. *Staying Alive: Women, Ecology and Development.* Zed Books, London.

Simmons, I.G. 1974. *The Ecology of Natural Resources.* Edward Arnold, London.

Simmons, I.G. 1979. *Biogeography: Natural and Cultural.* Edward Arnold, London.

Simmons, I.G. 1993. *Environmental History: A Concise Introduction.* Basil Blackwell Publishers, Oxford.

Simon, J. 1981. *The Ultimate Resource.* Martin Robertson, Oxford.

Simon, J.L. 1986. *Theory of Population and Economic Growth.* Basil Blackwell Oxford.

Singh, H.P. 1982. *Management of Desert Soils. Review of Soil Research in India*. 676 - 699, I.C.A.R., New Dehli.

Sivakumar, M.U.K. and Gnounou, F. 1987. Agroclimatology of West Africa: Burkina Faso. *International Crop Research Institute for the Semiarid Tropics* (I.C.R.I.S.A.T.), Patancheru.

Sivakumar, M.U.K., Piara - Singh, T. and Williams, J.H. 1987. Agro - Climatic Aspects in Planning for Improved Productivity of Alfisols in the Semiarid Tropics. *International Crop Research Institute for the Semiarid Tropics* (I.C.R.I.S.A.T), Patancheru.

Smith, R.L. 1995. *Ecology in Encyclopaedia Britannica* (Macropaedia), 14, 1069-1126. Encyclopaedia Britannica Inc., Chicago.

Soil Research Institute 1995. *Research Report.* SRI, Kumasi.

Solbrig, O.T. 1991. *Savanna Modelling for Global Change in Biology International (Special Issues)* 24, 1-47.

Solbrig, O.T. 1993. Ecological constraints to savanna land use. In Young, M.D. and Solbrig (Eds.) *The World's Savannas: Economic Driving Forces, Ecological Constraints and Policy Options for Sustainable Land Use,* 21-47. The Pantheon Publishing Group, Paris.

Soussan, J.G. 1992. Sustainable Development. In Mannion, A.M. and Bowlby, S.R. (Eds) 1992. *Environmental Issues in the 1990s*. John Wiley and Sons, Chichester.

Spaeth, H.J. 1996. *Land degradation in west Niger. Desertification Control Bulletin* (UNEP) 28: 34-41.

Sponsel, L.E., Headland, T.N. and Bailey, R.C. 1996. Tropical deforestation: the human dimension. *Biology and Resource Management in the Tropics* Series, Colombia University Press, New York.

Sprugel, D.G. 1991. Disturbance equilibrium and environmental variability: what is 'natural' vegetation in a changing environment. *Biological Conservation* 58: 1-8.

Squire, F.A. 1962. Mammalian Fauna. In Wills, J.B. 1962. *Agriculture and Land Use in Ghana,* 170-172, Oxford University Press, London.

Stamp, L.D. and Morgan, W.T.W. 1972. *Africa: A Study in Tropical Development* (3rd ed.). John Wiley and Sons, New York.

Steel, W. F, and Aryeetey, E. 1995. Informal savings collectors in Ghana. *Development and Cooporation* 16-17.

Stern, N. 1996. Growth theories, old and new: and the role of agriculture in economic development. *FAO Economic and Social Development Paper* (FAO).136, London School of Economics, London.

Stewart, F. 1995. *Adjustment and Poverty: Options and Choices.* Routledge, London.

Stock, W.D. and Allsopp, N. 1992. Functional Perspective of Ecosystems. In Cowling, R.M. 1992. *The Ecology of Fynbos: Nutrients, Fire and Diversity,* 241-259. Oxford University Press, Cape Town.

Stock, W.D., Van der Heyden, F. and Lewis, O.A.M. 1992. Plant structure and function. In Cowling, R.M. 1992. *The Ecology of Fynbos: Nutrients, Fire and Diversity,* 226-240. Oxford University Press, Cape Town.

Stocking, M.1993. *Soil Erosion in developing countries: where geomorphology fears to tread Discussion Paper No. 241*, Norwich School of Development Studies, University of East Anglia.

Stott, P.A. 1989. Lesson from an ancient land: Kakadu National Park, Northern Territory, Australia. *Plants Today* 2: 121-125.

Stott, P.A. 1990. Stability and stress in the savanna forest of mainland south East Asia. *Journal of Biogeography* 17: 373-383.

Stott, P.A. 1991. Recent trends in the ecology and management of the worlds's savanna formations. *Progress in Physical Geography* 15: 18-28.

Stott, P.A. 1994. Savanna landscapes and global environmental change. In Roberts, N. (Ed.) *The Changing Global Environment, 287-303.* Basil Blackwell Limited, Cambridge.

Stringer, E.T. 1972. *Foundations of Climatology: an Introduction to Physical, Dynamic, Synoptic and Geographical Climatology.* W.H. Freeman and Company, San Francisco.

Subrahmanian, R. 2005. Gender equality in education: Definitions and measurements. *International Journal of Educational Development* 25(4): 395-407.

Sullivan, S. 1996. Towards a nonequilibrium ecology: perspectives from an arid land. *Journal of Biogeography* 22: 1-5.

Sutton, M.Q. and Anderson, E.N. 2004. *Introduction to Cultural Ecology.* Altamira, Walnut Creek

Swaine, M.D. 1996. Rainfall and soil fertility as factors limiting forest species distributions. *Ghana Journal of Ecology* 84 (3): 419-428.

Sweetman, C. and Porter, F. (Eds). 2005. *Mainstreaming Gender in Development: A Critical Review.* Oxfam Focus on Gender Series. Oxfam, Oxford.

Tabor, S.R. 1995. *Agricultural Research in an Era of Adjustment: Policies, Institutions, and Progress.* World Bank, Washington DC.

Taddese, G. 2001. *Land degradation: a challenge to Ethiopia. Environmental Management* 27(6): 815 – 824.

Talbot, C. 1981. *Cited in Carson,* op cit p.45.

Tallet, B. 1985. Le Burkina Faso Espaces ethniques et migrations. Comment gerer le mouvement? *Polit Afr* 20, 65 - 77.

Tallis, J.H. 1991. *Plant Community History: Long Term Changes in Plant Distribution and Diversity*. Chapman and Hall, London.

Tansley, A.G. 1935. The use and abuse of vegetational concepts and terms. *Ecology* 16: 284-307.

Taylor, C.J. 1952. *The Vegetation Zones of the Gold Coast* Government Printing Department, Accra.

Technical Assistance to the Weija Irrigation Company (TAWEICO) Final Report 1986. Technosynesis SpA, Roma.

Tekle, K. and Hedlund, L. 2000. Land cover changes between 1958 and 1986 in Kalu District, Southern Wello, Ethiopia. *Mountain Research and Development* 20(1): 42-51.

Teme, B., Breman, H. and Sissoko, K. 1996. *Agricultural Intensification in the Sahel: myth or reality*? Institut d'Economie Rurale, Bamako.

The Jibowu Commission Report 1956. *Governement of the Gold Coast,* Accra.

Thebaud, B. and Batterbury, S. 2001. Sahel pastoralists: opportunists struggle conflict and negotiation. A case study from eastern Niger. *Global Environmental Change* 11: 69–78.

Thomas-Emeagwali, G. 1995. *Women Pay the Price: Structural Adjustment in Africa and the Carribbean*. Africa World Press, Trenton.

Thomas-Slayter, B.P. 1994. Structural change, power politics and community organisation in Africa: challenging the patterns, puzzles and paradoxes *World Development* 22(10): 1479-1490.

Thompson, H.N. 1910. Reports of Forests in the Gold Coast. Cited in Taylor, C. 1960. *Synecology and Silviculture in Ghana. Thomas* Nelson and Sons, London.

Tiffen, M., Mortimore, M. and Gichuki, F. 1994. *More people, Less erosion: Environmental Recovery in Kenya*. John Wiley and Sons, Chichester.

Timaeus, I.M. and Onuoha, N.C. 1995. Has a fertility transition begun in West Africa? *Journal of International Development* 7 (1): 93-116.

Tisdell, C.A. 1991. *Economics of Environmental Conservation: Economics for Environmental Ecological Management*. Elsevier, London.

Tivy, J. and O'Hare, G. 1981. *Human Impact on the Ecosystem.* Oliver and Boyd, Edinburgh.

Todaro, M.P. 1989. *Economic Development in the Third World* (4th. ed.). Longman, London.

Tognetti, S.S. 1995. *Poverty and the Environment: Reconciling Short Term Needs with Long Term Sustainability Goals*. United Nations Environment Programme, Nairobi.

Tonah, S. 1994. Agricultural extension services and smallholder farmer's indebtedness in Northeastern Ghana. *Journal of Asian and African Studies* 29(1-2): 119-128.

Toulmin, C. 1990. *Natural Resource Management in the Sahel: Bridging the Gap Between Bottom Up and Top Down.* Workshop on Human Life in Arid Lands, Uppsala.

Trainer, T. 2010a. *The Transition to a Sustainable and Just World.* Envirobook, Sydney.

Trainer, T. 2010b. De-growth is not enough. *International Journal of Inclusive Democracy.*6, 4,

Tribe, M. 1995. Rural small-scale industrial development in Ghana: an exploration of issues. *Science, Technology and Development* 13 (1): 50 - 62.

Troeh, F.R., Hobbs, J.A. and Donahue, R.L. 1980. *Soil and water conservation for productivity and environmental protection.* 25 -51: 454 - 487, Prentice Hall Inc., Englewood Cliffs.

Tsey, K., Lubrani, O., Schmid-Hageth, A. 1995. The role of indigenous social organisations in rural development in a West African village: some lessons for NGOs. *Development Bulletin* 35: 47-57.

Tudge, C. 1991. *Global Ecology.* Natural History Museum Publications, London.

Turner, B.L. 1989. The specialist synthesis approach to the revival of geography: the case of cultural ecology. *Annals of the Association of American Geographers* 79 (1): 88-100.

Turner, T.E. and Brownhill, L. 2006. Ecofeminism as gendered, ethnicized class struggle. *Capitalism Nature Socialism. 17*(4): 87-95.

United Nations 2012. *Geographical region and composition URL:* http://millenniumindicators.un.org/unsd/methods/m49/m49regin.htm

UNICEF 1986. *Adjustment Policies and Programmes to Protect Children and Other Vulnerable Groups UNICEF*, Accra.

UNICEF 1990. *Children and Women of Ghana: A Situational Analysis UNICEF,* Accra.

United Nations Convention to Combat Desertification 2002. Republic of Ghana Second National Report to the Conference of Parties to the United Nations Convention to Combat Desertification URL: http://wwwunccdint/cop/reports / africa/national/2002/ghana-engpdf

Van Niekerk, R. V. 1995. *The state of agricultural information in Africa. Agricultural Information in Africa 57-66,* Third World Information Services, Ibadan.

Van Warmelo, N.J. 1962. *Notes on the Kaokoveld (South West Africa).* Cape Times Limited, Cape Town.

Vayda, A.P. and Walters, B.B. 1999. *Against political ecology.* Human Ecology 27(1): 167-179.

Vlassoff, C. 1994. From rags to riches: the impact of rural development on women's status in an Indian village *World Development* 22(5): 707-719.

Volta, B. 1971. The Future of Ghana's wildlife *The Ghana Farmer* XV(1): 30-35.

Walker, P.A. 2005. Political ecology: where is the ecology? *Progress in Human Geography* 29(1):73–82.

Walker, P.A. 2006. Political ecology: where is the policy*? Progress in Human Geography* 30(3): 382-395.

Wallace, I. 1997. Agricultural Education at the crossroads: present dilemmas and possible options for the future in sub-Saharan Africa. *International Journal of Educational Development* 17 (1):27-39.

Wane Conde, C. 1985. *Enquetes et reflexions sur des projects de reboisement a paysanne dans quatre pays saheliens:* Burkina Faso, Mali, Niger, Senegal. Assoc Bois de Feu - Fonds d' Aide et de Cooperation, Paris.

Warner, M.W. and Balcombe, K. 1996. How Useful in Gender Roles of Africa's Rural Peoples? *An Application in Logistic Regression Analysis.* Wye College Press, Ashford.

Warner, M.W., Al-Hassan, R.M. and Kydd, J.G. 1997. Beyond gender roles? Conceptualizing the social and economic lives of rural peoples in sub-Saharan Africa. *Development and Change* 28(1):143-168.

Watts, I.E.M. 1955. *Equatorial Weather.* University of London Press, London.

Watts, M. 2009. *The Dictionary of Human Geography.* John Wiley & Sons, Chichester

Webster, C.C. and Wilson, P.N. 1966. *Agriculture in the Tropics,* Longman Green and Company, London.

Weija Irrigation Company 1993. Annual Report Government Printer, Accra.

Weija Irrigation Company 1994. Annual Report Government Printer, Accra.

Weija Irrigation Company 1995. Annual Report Government Printer, Accra.

Weija Irrigation Company 1996. Quarterly Report Government Printer, Accra.

Weins, J. 1984. On understanding a non-equilibrium world: myth and reality in community patterns and processes. In Strong, D.R. (Ed.) *Ecological*

Communities: Conceptual Issues and the Evidence, 439-457. Princeton University Press, Princeton.

Weiss, J. 1995. *Economic Policy in Developing Countries: The Reform Agenda.* Prentice Hall, London.

Wellens, J. and Millington, A.C. 1992. Desertification. In Mannion A.M. and Bowlby, S.R. (Eds.) *Environmental Issues in the 1990s,* 245 - 261. John Wiley and Sons, Chichester.

Werner, P.A. 1990. Ecological determinants of savannas: abiotic and biotic. *Journal of Biogeography* 17: 401-402.

Werner, P.A., Walker, B.H. and Stott, P.A. 1990. *Introduction. Journal of Biogeography* 17: 343-344.

West, L.T., Wilding, L.P. and Calhoun, F.G. 1987. Argillic horizons in sandy soils of the Sahel, West Africa. In Fedoroff, N., Bresson L.M. and Courty, M.A. (Eds). *Soil Micromorphology,* 221 -225. Association Francaise pour l Etude du sol Versailles.

Westoby, M., Walker, B. and Noy-Meir, I. 1989. Opportunistic management for rangelands not at equilibrium. *Journal of Range Management* 42: 264-274.

Whittaker R.H. 1953. A consideration of climax theory: the climax as a population and pattern. *Ecological Monographs,* 23: 41-78.

Whyte, A.V.T. 1986. From hazard perception to human ecology. In Kates, R.W. and Burton, I.(Eds) Geography, *Resources and Environment* 2, 240-271, University of Chicago, Chicago.

Wickham, T. 1986. Irrigation systems updating, planning and management. *Span* 29(1): 15-17.

Wijlkman, A. and Timberlake, L. 1986. Is the African drought an act of God or man ? *The Ecologist.,* 15 (112): 9 - 18.

Wilshusen, P.R. 2003. Exploring the political contours of conservation: a conceptual view of power in practice. In Brechin, S.R., Wilshusen, P.R., Fortwangler, C.L. and West, P.C. (Eds), *Contested Nature,* 41-57. University of New York Press, Albany.

Williams, A.R. 1957. Cattle droving in Ghana. *Bulletin of the Ghana Geographical Association* 2(2): 6-11.

Witcombe, J.R. and Joshi, A. 1996. Farmer Participatory Crop Improvement: 2. Participatory Varietal Selection, a Case Study in India. *Experimental Agriculture* 32 (4):461-477.

Witcombe, J.R., Sthapit, B.R. and Joshi, K.D. 1996. Farmer participatory crop improvement 3. participatory plant breeding, a case study for rice in Nepal. *Experimental Agriculture* 32 (4): 481-496.

Wolf, E. 1972. Ownership and political ecology, *Anthropological Quarterly* 45(3): 201-205.

Wolf, E. 1997. *Europe and the People Without History.* University of California Press, Berkeley and Los Angeles.

Woodgate, G. 1991. Agroecological possibilities and organisational limits: some initial impressions from a Mexican case study. In Goodman, D. and Redclift, M. (eds.) *Environment and Development in Latin America: The Politics of Sustainability,* 155-183. Manchester University Press, Manchester.

Woodgate, G.R. 1992. *Sustainability and the Fate of the Peasantry: the Political Ecology of Livelihoods Systems in an Upland Agroecosystem in the Central Highlands of Mexico.* Unpublished Ph.D Thesis, University of London.

Worch, B., Kappel, R. and Farber, H. 1989. Africa's Ecological Disaster: An Introduction. In *African Perspectives Year Book* (Volume 1: Human Dimensions of Adjustment), 251 - 256, Verlag Schelzky and Jeep, Berlin.

World Bank 1989. *Ghana: Structural Adjustment for Growth.* World Bank, Accra.

World Commission on Environment and Development 1987. *Our Common Future.* Oxford University Press, Oxford.

Yang, D.Y. 1997. Education and off-farm work. *Economic Development and Cultural Change* 45 (3): 613-632.

Yayah, R.T. 1989. *Ghana Seminar on Forestry Statistics in Africa.* Ghana Goverment, Accra.

Yearley, S. 1991.*The Green Case: a Sociology of Environmental Issues, Arguments and Politics.* Harper Collins, London.

Yearley, S. 1994. Social movements and environmental change. In Redclift M. and Benton, T. 1994. *Social Theory and the Global Environment*, 150-168. Routledge, London.

Zanen, S. 1996. Development: Idea and Action: Confrontation of Farmer and Donor Strategies in the Practice of Development Co-Operation (the Netherlands-Burkina Faso). *Centre for Non-Western Studies* (CNWS) Publications, Leiden University.

Zelezny, L, Poh-Pheng Chua, P. and Aldrich, C. 2000. Elaborating on gender differences in environmentalism. *Journal of Social Issues.* 56: 443-457.

Zimmerer, K.S. 1994. Human geography and the 'new ecology': the prospect of integration. *Annals of the Association of American Geographers* 84 (1): 108-125.

Zimmerer, K. 1996. Ecology as cornerstone and chimera in human geography. In Earle, C., Mathewson K. and Kenzer M. (Eds.) *Concepts in human geography,* 161-188. Rowman and Littlefield Publishers, Lanham.

Zinyama, L.M. 1995. Sustainability of smallholder food production systems in southern Africa: the case of Zimbabwe. In Binns, T. (Ed.) *People and Environment in Africa,* 215 - 224. John Wiley and Sons, Chichester.

Zonn, S.V. 1986. *Tropical and Subtropical Soils Science.* Mir Publishers, Moscow.

Zwarteveen, M.Z. 1994. *Gender Issues, Water Issues: a Gender Perspective to Irrigation Management.* International Irrigation Management Institute, Colombo.

Zweifel, H. 1995. Modern biotechnologies in agriculture: impact on women in the South. *Biotechnology and Development Monitor* 23, 10-13.

APPENDICES

APPENDIX 1. ECOLOGICAL SURVEY RESULTS

Vegetation	Species characteristics
Tree	Average of 23 per $100m^2$ quadrat commonest species being neem *Azadirachta indica* and mango Mangifera indica. Others included baobab *Adansonia digita*, coconut Cocus nucifera L., raintree *Samanea saman* and silk cotton *Ceiba pentandra.*
Shrubs	Average of 13 per $100m^2$ quadrat commonest species being neem *Azadirachta indica*. Others included *Fagara zanthoxlyloides*, *Milletia thonningi* and *Baphia nitida*.
Grass	Average of three species per quadrat. Commonest species *Andropogon*, *Sporobolus*, and *Panicum*.

Vegetation Change in the Study Area, by Percentage of Photo Coverage (1960 – 1996)

Areas	Date	Tree/shrub (%)	Grass/farm (%)	Settled areas (%)
Tuba	1960	17.3	81.6	1.1
	1996	12.6	86.2	1.2
Aplaku	1960	60.9	38.2	0.9
	1996	25.7	73.4	0.9
Bortianor	1960	17.2	78	4.8
	1996	9.1	77.1	13.8
Kokrobite	1960	45.5	53.2	1.3
	1996	51.6	46.2	2.2
Oshiyie	1960	18.7	79.8	1.5
	1996	20.0	78.1	1.9
Hinterland	1960	98.2	1.8	0
	1996	39.5	60.5	0

APPENDIX 2. SOILS OF THE STUDY AREA

The classification of the soils in the study area is shown in the diagram. Next shows a comparison of the local, FAO-UNESCO and USDA classification systems (Ahn 1995; Soil research Institute 1995).

Orders	Suborders	Great Groups
Alfisols	Ustalfs	Haplustalfs
	Aqualfs	Natraqualfs
Entisols	Psamments	Ustipsamments
	Orthents	Ustorthents
	Aquents	Endoaquents
Ultisols	Ustults	Haplustults
	Aquults	Paleaquults
Inceptisols	Aquepts	Endoaquepts

Local Classification	FAO - UNESCO Classification	USDA Classification
Fete Bediesi Complex		
Fete Series	Haplic Lixisols	Haplustalfs
Bediesi Series	Dystric Nitisols	Paleaquults
Oyarifa Consociation		
Mamfe Series	Ferric Acrisols	Haplustults
Korle Series	Dystric Leptosols	Ustorthents
Salom Series	Haplic Lixisols	Haplustalfs
Kweman Series	Ferric Acrisol/D. Regosol	Haplustalfs/Ustorthent
Krabo Series	Dystric Gleysols	Endoaquepts
Hacho Series	Dystric Regosols	Ustorthents
Nyibenya - Hacho Complex		
Nyigbenya Series	Ferric Acrisols	Haplustults
Toje Series	Eutric Regosols	Ustorthent
Korle Series	Dystric Leptosols	Ustorthents
Fete Series	Haplic Lixisols	Haplustalfs
Adentan Series	Ferric Lixsols	Haplustalfs
Hacho Series	Dystric Regosols	Ustorthents
Papao Series	Eutric Gleysols	Endoquents
Korle Consociation		
Korle Series	Dystric Leptosols	Ustorthents
Nyigbenya Series	Ferric Acrisols	Haplustuults
Mamfe Series	Ferric Acrisols	Haplustuults

Local Classification	**FAO - UNESCO Classification**	**USDA Classification**
Adentan Series	Ferric Lixisols	Hpalustalfs
Keta - Oyibi Association		
Keta Series	Eutric Regosols	Ustorthents
Oyibi Series	Sodic Solonchaks	Endoaquents
Truku Series	Gleyic Solonets	Natraqualfs
Muni Series	Sodic Solonchaks	Endoaquents
Adawso - Bawjiasi/Nta Ofin Compound		
Adawso Series	Haplic Lixisols	Haplustalfs
Bawjiasi Series	Haplic Lixisols	Haplustalfs
Nta Series	Gleyic Arenosols	Ustipsamments
Ofin Series	Dystric Gleysols	Endoaquents

Soil Type Linked to Elevation and Slope

Soils associations	Slope $^{o)}$)	Elevation (m)	Characteristics
Fete Bediesi Complex: Haplustalfs, Paleaquults	15.5	62	Pale, sandy clays and alluvial sands
Oyarifa Consociation: Haplustults, Ustorthents, Haplustalfs, Endoaquepts,	13.6	115	Reddish/pale sandy clays and poorly drained loams
Nyibenya Hacho Complex: Haplustults, Ustorthents, Haplustalfs, Endoquents	8	50	Well drained sands, and poorly drained clays
Korle Consociation: Ustorthents, Haplustuults, Haplustalfs	16.9	129	Reddish drained clays and heavy grey clays
Oyibi-Muni Association: Endoaquents, Natraqualfs	4.9	9	Heavy blue clays and grey sands
Songaw Consociation: Endoaquents,	4.7	10	Heavy calcareous and grey clays
Adawaso-Bawjiasi/Nta-Ofin Consociation: Haplustalfs, Ustipsamments, Endoaquents	2.8	15	Brown concretionary clays and red loams
Avieme-Zipa Association:	11.7	13	Red and yellowish sands and sandy clays
Keta and Keta-Oyibi Association: Ustorthents, Endoaquents	2.4	5	Pale grey brown loose sands

Source: Soil Research Institute 1995 and Field Survey

Details of the Soil Types

Alfisols

These soils occur largely under grasslands within 10^0 degrees of the equator, with a distinct dry season, and parent materials high in quartz and lower in iron. Alfisols usually possess an argillic B horizon, a base saturation of over 35 percent, a gravelly horizon at shallow depths, a light textured surface horizon and a high proportion of kaolinitic clays. There is often rock within 100 centimetres of the surface, which provides a source of weatherable nutrients within reach of many plant's roots. The fertility of these soils is low. The kaolinitic clays have a low pH, and hence a low cation exchange capacity. This inhibits nutrient absorption by plants. Concretions of weathered iron also form in the profile. Also monsoon rains cause erosion (Foth and Schafer 1980; Brady 1984; Ahn 1993; Lal 1993). The alfisols of the Accra-Ho-Keta plains, also termed the tropical grey earths, are grey soils developed over acid parent rocks. There is frequently a hard, sandy clay pan several centimetres below the surface. The low nitrogen, phosphorus and organic matter renders them infertile, and therefore they are rarely cultivated. (Dickson and Benneh 1988). The table below shows the alfisols of the study area.

Alfisols of the Study Area

Haplustalfs	Fete Series, Salom Series, Adentan Series, Adawso Series, Bawjiasi Series
Natraqualfs	Truku series

Source: Soil Research Institutes (1995).

1. Haplustalfs

(a) Fete Series

The fete series are well drained, pale brown soils, which developed on steep slopes. They contain a little weathered brash of Togo quartzite, or rarely, sandstone in the profile at a shallow depth. Usually the profile consists of a shallow, sandy humus topsoil. There may be small rock fragments. This topsoil merges into pale, grey brown to pale yellow brown sand, sandy light clay, or sand. This subsoil also has rock fragments. The soils of the fete series have low fertility. For natural restorations of fertility, a 10 year fallow period is needed. Mechanised agriculture is not easy on these soils, due to the relevance of rock fragments and steep slopes (Brammer 1967; Adu and Ascainah 1992; Soil Research Institute 1995).

(b) Salom Series

The soils of this series are similar to those of the fete series. In many areas they are found together. The salom series are however developed over Togo phyllites, shales and serricite schist. Their texture is thus clayey rather than sandy. They commonly occur on slopes. The topsoils are dark brown to dark grey brown, humus friable, loam to light clay. There are frequent occurrences of gravel and sand which are washed in from associated soils. The subsoils are paler and moderately clayey. This merges at 6 to 12 metres with brashy weathered rock. The subsoils exhibit slight variations in accordance with the nature of the weathered parent rock. Where this is clay shale, the subsoil is pale grey brown to pale yellow brown (Brammer 1967). Where the parent rock is serricite schist, (which is quite rare) the colour varies from yellow to pale red. The texture then tends to be friable clay. Where the parent rock is phyllite (the commonest), the subsoils are similar in texture to the last group, but pale orange brown to pale red. These soils are generally well drained even though there are clayey horizons (Brammer 1967; Adu and Asiama 1992; Soil Research Institute 1995).

(c) Adentan Series

This series consists of soils deposited on gentle slopes. The parent material is slope wash from gravelly and concretionary upper slope soils. These soils often lie between red, upper slope soils and loose sandy, pale coloured lower slope and valley soils. The topsoils tend to be orange brown to yellow brown sandy loams to sandy clays. These grade into subsoils of mottled sandy clay, with abundant ironstone concretions and quartz gravel. They are underlain by weathered crystalline rock. These soils are readily drained. This means that the retentive capacity of the soil is low, causing serious desiccation during the dry season. The clay content of these soils tends to be low and the fertility is also low, limited to the thin topsoil. For intensive cultivation, fertilisers are needed (Brammer 1967; Adu and Asiamah 1992; Soil Research Institute 1995).

(d) Adawso Series

This series consists of grey soils developed over little weathered, biotite granite with veins of pegmatite. The topsoil consists of dark greyish brown, slightly humus sandy light loams, grading to brown loamy sands. These are weak, fine, granular and loose, with a few small quartz stones. The interface between the topsoils and the subsoils consists of yellow brown loamy sand, very weak fine granular and loose, with occasional gravel, ironstone nodules and quartz stones. The subsoils grade from: brown, gritty clay loams with quartz stones, gravel and iron modules; through reddish yellow sandy clay with rare ironstones and decomposed granite, and common fine quartz gravel; to reddish yellow and mottled yellow sandy clays, fine and firm with occasional ironstone concretions and decomposed granite. Decomposed granite is common at the interface of the weathered substratum (Brammer 1967; Adu and Asiamah 1992; Soil Research Institute 1995).

(e) Bawjiasi Series

Like the Adawso series, these are pale yellow brown and reddish brown gravelly concretionary clays, developed over slightly weathered, biotite granite on summits and upper slopes. The topsoil grades from dark reddish brown loams, to sandy clay loams to reddish brown gritty clay loams. There are rare occurrences of quartz gravel. The subsoil grades from reddish brown gritty fine firm clays with frequent quartz gravel; to yellowish red gritty clays with very frequent quartz gravel and stones; and in some cases with decomposed biotite granite. The subsoil layer over the weathered substratum consists of reddish yellow sandy loams. There are traces of decomposed biotite granite. Both the Adawso and Bawjiasi series tend to be suitable for food cropping. Nevertheless, fertilisers are be necessary. (Brammer 1967; Adu and Asiama 1992; Soil Research Institute 1995)

2. Natraqualfs

(a) Truku Series

This series consists of silty, blue grey alluvial clays overlain by a few centimetres of black mud. In many areas these soils are subjected to seasonal flooding. When dry, they develop reddish surface layers, with strongly mottled horizons. They are not deemed to be important agriculturally (Brammer 1967; Adu and Asiamah 1992; Soil Research Institute 1995).

Entisols

These are recent soils with under-developed genetic horizons. There may be an ochric surface horizon, but the main feature is the absence of profile development. This is because in general they have been deposited too recently. In many cases there is no genetic horizon in the subsoil. Entisols include rocky shallow soils on steep slopes. These develop may over resistant parent materials (known as lithosols) or over quartz sands (known as tegosols).

Entisols of the Study Area

Ustorthents	Hacho Series, Toje Series, Korle Series, Keta Series.
Ustipsamments	Doyom Series, Nta Series
Endoaquents	Papao Series, Oyibi series, Ofin Series.

Source: Soil Research Institute (1995).

1. Ustothents

(a) Hacho Series

This series is formed from poorly drained, lower slope, slope wash material. The topsoil is humus stained. The subsoil may be slightly mottled orange. The soil profile consists of a few metres of pale brown sand, which grades to sandy clay with depth. Where the soils lie under natural vegetation, they are well drained. However, where the soils are exposed, the surface readily becomes impermeable. Thus, some areas become water logged during the rainy season. However, as these soils have a low water retentive capacity, they are subject to desiccation during the dry season. The nutrient content of Hacho soils is low. The topsoils are slightly acid and the deeper layers increase in acidity. Therefore, heavy fertilisation is required, for the cultivation of the soil (Brammer 1967; Adu and Asiamah 1992; Soil Research Institute 1995).

(b) Toje Series

The soils of this series formed over Togo quartzite schists, sandstones and Accraian shales. The top soils grade from dark brown to brown, slightly humus crumbly, loose, fine sand; to reddish brown slightly humus crumbly loose fine sand; to weak red, firm, porous structured fine sands. The subsoils are red, firm porous, structureless, light loams grading to red very firm porous rather gritty light loams. These soils are well drained, but absorb water best when there is plant cover, minimising erosion and run off. They are slightly acid, with low nutrient reserves. The top soil organic matter is the main holder of nutrients. Nitrogen and phosphorus are low in occurrence (Brammer 1967; Adu and Asiamah 1992; Soil Research Institute 1995).

(c) Korle Series

This series is composed of well drained red, light clays, on slopes. Brashy Togo quartzite schist occurs at shallow depths in the profile. The soils are moderately fertile. The topsoils are almost neutral but as their depth increases acidity becomes pronounced. Locally, slope wash from the upper slopes may accumulate in low lying areas. This leaves orange brown to yellow brown, gravelly loams and clays. There is commonly excessive drainage, and therefore during the dry season the soils become very desiccated. This series usually occurs under thickets or grassland; on moderate slopes of small hills (Brammer 1967; Adu and Asiamah 1992; Soil Research Council 1995).

(d) Keta Series.

This series lies to the west of the Songaw lagoon. The coastal sand dunes occur with saline and brackish water lagoons. The soils consist of a top layer of 7 cm of dark grey to pale brown humus stained loose sand, and a subsoil of pale yellow sand which is weakly mottled orange, with shells and shell fragments. (Brammer 1967; Adu amd Asiamah 1992; Soil Research Institute 1995).

2. Ustipsaments

(a) Doyum Series

This consists of loose, course deep, pale coloured sands. They are developed from weathered granite on upper slopes, and undulating slopes. They are similar to the soils of the Hacho series, but minor differences occur. Doyum soils tend to occupy upper slopes rather than lower slopes, and they tend to be composed of course rather than medium or fine sand. The surface 30 centimetres is stained by humus, overlying 120 to 150 centimetres of pale yellow course sand. There is usually a thin layer of strongly mottled, gritty, gravelly clay at the base of the profile, this occurring over weathered mottled granite.

In local areas, orange, or orange and black iron pan may replace the mottled clay layer. Both this and the mottled clay can serve as impediments to drainage during the dry season. The topsoil and upper subsoil are however well drained and desiccated during the dry season. The impermeable layers may create temporary shallow water tables and local surface waterlogging during the rainy season. The profile of the Doyum series is neutral to slightly acid in reaction. The nutrient supply and organic content of the whole profile, including the topsoil is low (Brammer 1967; Adu and Asiamah 1992; Soil Research Institute 1995).

(b) Nta Series

These soils are deposits of medium to course sands, which occur on the lower slopes and drainage beds of undulating topography. The profile which may vary in depth from 60 to 300 centimetres, overlies partially weathered granite. In some instances a course stone line separates the main profile from the underlying weathered granite. The profile is largely imperfectly drained, pale yellow brown, sandy soils with frequent occurrences of ironstone in or immediately above the weathered granite. The topsoil has a texture of loamy sand to sandy loam and/or gritty loam. The subsoil is loamy sand to sandy loam. For a detailed description; the topsoil consists of very dark greyish brown, slightly humus single grained and loose loamy sand, grading into dark brown single grained loamy sand. The subsoil grades from yellowish brown, single grained and loose loamy sand, to pale yellow, single grained fine sand, to pale single loose fine sand. There may be occasional irregularly shaped, small and medium-seepage ironstone concretions; and low in the profile, abundant small and large subangular and quartz gravel. In some instances a water table exists at this level, just above the weathered granite substratum (Brammer 1967; Adu and Asiamah 1992; Soil Research Institute 1995).

3. Endoaquents

(a) Papao Series

These soils are poorly drained, grey brown, silty clays occurring mainly in valley bottoms. The topsoil is a dark grey brown, humus porous, silty loam. This layer is powdery but the layer immediately below is firm when dry, becoming sticky when moist. The subsoil consists of 1 to 2 metres of grey brown, porous, silty light clay, weakly mottled. This is penetratable when wet. However, when dry they are nearly indurated and very hard. There are also soft red ironstone and black manganese concretions occurring in the profile. These soils are distinctive in that when dry their colour is whitish grey. During the dry season the soils dry out due to their poor retentive capacity, but during the rainy season the soils are waterlogged for long periods, with local flooding being prevalent. The papao series is moderately acid. The nutrient capacity is limited largely to the topsoil. For cultivation the soil requires strong fertiliser application. Also mechanised tillage often brings clay to the surface causing the formation of hard clods when dry (Brammer 1967; Adu amd Asiamah 1992; Soil Research Institute 1995).

(b) Oyibi Series

This occurs in association with the keta series as part of the Keta-Oyibi Association. It is composed of grey mottled orange or heavy red clays. These are periodically flooded by fresh or saline water, and support sedges or marsh grasses. In most instances there is dark to pale grey, mottled and black heavy clay. In a few profiles, there may be a topsoil of grey brown sand or sandy loam instead of the heavy clay. In the commoner heavy clay profile, gypsum crystals may occur in areas of periodic waterlogging. Sandy horizons may occur below 75 centimetres. There may also be buried mangrove peat or black humus occurrences in the profile (Soil Research Institute 1995). The poor drainage of these soils allows the development of moderately deep water tables of brackish or saline water. (Brammer 1967) notes that for cultivation, there should be drainage to leach the sodium (Na). Lime must be applied to combat the acid reaction. The problem to be solved would then be their sticky nature when wet, and the compact and hard structure when dry.

(c) Muni Series

These soils comprise alluvial lagoon clays. When dry they form salt flats around lagoon margins. When moist they are flooded by saline water. The topsoil is composed of 5 to 10 centimetres of grey or yellow grey sand or sandy clay with surface crusts of salt. The subsoils are composed of alluvial layers of greyish sand and clay, which is usually mottled black and reddish purple. Shells occur frequently in the profile. Due to their high saline content, these soils are not agriculturally developed (Brammer 1967; Adu and Asiamah 1992; Soil Research Institute 1995).

(d) Offin Series

These soils are developed from grey alluvial sands, a few metres thick, which overlie successively stream gravel and decomposed bedrock. The topsoils vary in texture from sand to loam. The subsoils are sandy with exceptional profiles containing stratified layers of clay and sand. During the wet season, these soils are waterlogged and in extreme cases flooded. During the dry season there may be pronounced desiccation, especially after clearance of vegetation cover. The Offin soils are similar to the Nta soils in that they are very low in nutrients (Brammer 1967; Adu and Asiamah 1992; Soil Research Institute 1995).

Ultisols

These soils (sometimes termed the forest ochrosols) are the main soils under the moist semi deciduous forest. They are highly weathered, but not to the same extent as oxisols. This is largely due to the rainfall intensity. Variations in these soils reflect the length of the dry seasons. The general fertility of these soils is low. This is because the dominant mineral, kaolinite, has a low cation exchange capacity, and the cations absorbable by plants as nutrients are few. Calcium, magnesium and potassium are readily leached out of the profile. Exchangeable aluminium accumulates, and this may be toxic to some plants. The subsoil has poor permeability and is prone to waterlogging and erosion (Ahn 1993).

Ultisols of the Study Area

Paleaquults	Bediesi Series
Haplustuults	Mamfe Series

Source: Soil Research Institute (1995).

1. Paleaqualts

(a) Bediesi Series

The profile of these soils consists of a surface layer of approximately 18 centimetres of dark brown to brown sandy loam, overlying approximately 20 centimetres of reddish brown sandy clay loam to clay. There are occasional occurrences of quartz gravel and ferruginised sandstone boulders. This overlies a mottled orange and brown sandy loam layer which is approximately 20 centimetres thick. Beneath this is a layer of ferruginised and indurated brashy loam, which grades into clay shale rocks and sandstone boulders. (Brammer 1967; Adu and Asiamah 1992; Soil Research Institute 1995).

(b) Mamfe Series

The profiles of these soils are overlain by topsoils of very dark greyish brown, loose sandy light loams, grading to dark brown, sandy light loams with occasional ironstone concretions and ironstained quartzite. The subsoils grade from brown sandy clay, with frequent ironstone concretions, weathered phyllite, and quartzite sandstone, to reddish brown sandy clay with abundant ironstone concretions and rare stones and gravel of ferruginised rocks, to pale red and yellow sandy clay with occasional ironstone concretions, usually with interstices of disintegrated ferruginised quartzite. The weathered substratum consists of pale yellow ferruginised quartzite (Brammer 1967; Adu and Asiamah 1992; Soil Research Institute 1995).

Inceptisols

These soils are also termed the groundwater laterites. The name `inceptisols' comes from `inception' which means beginning. They occur mainly in parts of the interior wooded savanna. They are recent soils, but they have more profile development than the entisols. There is usually a diagnostic subsoil horizon (a cambic horizon). A common feature of these soils is the prevalence of an almost impermeable layer of iron pan (cemented ironstone) causing waterlogging at the surface during the rainy season. The permeability of iron pan layer determines the extent of waterlogging. The nutrient level and organic matter content is low. The acidity is largely due to the erosion and leaching (Dickson and Bemreh 1988).

Inceptisols of the Study Area

Endoaquepts	Krabo Series

Source: Soil Research Institute (1995).

1. Endoaquepts

(a) Krabo Series

These soils comprise grey brown, silty clays. The topsoil is a dark brown, humus, silty loam, which is generally firm and porous. The topsoils are powdery when dry and sticky when wet. Ironstaining may also occur, causing the soil to be mottled with yellow or brown. When dry, the soil is hard, nearly indurated and white. In texture, the soil is silty loam for the first 30 to 60 centimetres, grading into clayey or silty clay which continues to a depth of 180 centimetres. There may be iron pan in the profile. Due to their clayey nature, these soils are waterlogged during the wet season. During the dry season however they may be deeply desiccated. The bulk of the soils nutrient capacity is located in the topsoil. The soils range in pH, the majority occupying a range between neutral and moderately acid. The soils are suitable for cultivation, but fertilisers, especially phosphorus (P) and nitrogen (N), are required (Brammer 1967; Adu and Asiamah 1992; Soil Research Institute 1995).

APPENDIX 3. SOIL EROSION METHODOLOGY

No. of Quadrats	**1,350**
Area of Quadrats	4 sq.m
Vegetable Type	
Short Fallow Quadrats	750
Grass	
Average Ground Exposure (%)	40
Shrubs	
Average Ground Exposure (%)	5
Mixed Vegetation	
Average Ground Exposure (%)	20
Long Fallow Quadrats	600
Grasses	
Average Ground Exposure (%)	5
Shrubs	
Average Ground Exposure (%)	2
Mixed	
Average Ground Exposure (%)	3

Soil Erosion under Fallow Vegetation

Variable	**Short Fallow**	**% area**	**Long Fallow**	**% area**
	Main Erosion Type		**Main Erosion Type**	
Grasses				
Slope				
0	Sheet/rill	30	None	0
5	Sheet/rill	30	None	0
10	Sheet/rill	35	Sheet and rill	5
15	Sheet/rill/gulley	40	Sheet and rill	10
20	Sheet/rill/gulley	45	Sheet and rill	10
25	Sheet/rill/gulley	50	Sheet and rill	20
Shrubs				

(Continued)

Variable	Short Fallow	% area	Long Fallow	% area
	Main Erosion Type		Main Erosion Type	
Slope				
0^0	Sheet	2	None	0
5^0	Sheet	5	None	0
10^0	Sheet and rill	10	Sheet and rill	5
15^0	Sheet/rill/gulley	15	Sheet and rill	5
20^0	Sheet/rill/gulley	16	Sheet/rill/gulley	8
25^0	Sheet/rill/gulley	18	Sheet/rill/gulley	10
Grass/shrub				
Slope				
0^0	Sheet and rill	15	None	0
5^0	Sheet and rill	15	None	0
10^0	Sheet/rill/gulley	18	Sheet and rill	5
15^0	Sheet/rill/gulley	25	Sheet/rill/gulley	10
20^0	Sheet/rill/gulley	30	Sheet/rill/gulley	16
25^0	Sheet/rill/gulley	38	Sheet/rill/gulley	18

Appendix 4. Desiccation Prone Soils in the Study Area

Soil Type	**Texture**	**Main location**
Haplustalfs		
Fete Series	Sandy clay/sand	Steep slopes
Adentan Series	Sandy loam/sandy clay	Mid slopes/Upper slopes
Adawso Series	Sandy light loam/loamy sand	Mid slopes
Ustorthents		
Toje Series	Fine sand/light loam	Mid slopes
Korle Series	Light clays/ some gravel	Mid slopes
Ustipsamments		
Doyum Series	Coarse sand/iron pan	Upper slopes
Endoaquents		
Papao Series	Silty loam/silty clay	Valleys

Soil Type	Texture	Main location
Ofin Series	Sandy loam/sand	Valleys
Paleaquults		
Bediesi Series	Sandy loam/sandy clay loam	Mid slopes
Haplustults		
Mamfe Series	Sandy light loam/clay - ironstone	Mid slopes
Endoaquepts		
Krabo Series	Silty clays/silty loams	Mid slopes/valleys

Source: Field Survey, Brammer (1967) Adu and Asiamah (1992,) SRI Report (1995).

Crop Losses to Desiccation (1965 - 1987)

Year	% loss (hills)	% loss (plains)	Main losses	Lowest losses
1965	10	2	Tomatoes	Maize
1966	10	2	Tomatoes	Maize
1967	11	2	Tomatoes	Maize
1968	11	2	Tomatoes	Cassava
1969	12	2	Tomatoes	Maize
1970	12	5	Tomatoes	Maize
1971	12	5	Tomatoes	Cassava
1972	15	8	Tomatoes	Cassava
1973	17	8	Tomatoes	Maize
1974	20	10	Tomatoes	Cassava
1975	23	10	Tomatoes	Maize
1976	22	10	Tomatoes	Cassava
1977	24	10	Tomatoes	Maize
1978	26	15	Tomatoes	Cassava
1979	30	15	Tomatoes	Cassava
1980	32	16	Tomatoes	Cassava
1981	35	17	Tomatoes	Cassava
1982	35	16	Tomatoes	Cassava
1983	35	16	Tomatoes	Maize
1984	38	17	Tomatoes	Maize
1985	40	17	Tomatoes	Maize
1986	42	17	Tomatoes	Cassava
1987	44	18	Tomatoes	Maize

(Continued)

Year	% loss (hills)	% loss (plains)	Main losses	Lowest losses
1988	45	18	Tomatoes	Maize
1989	46	20	Tomatoes	Cassava
1990	50	20	Tomatoes	Maize
1991	50	20	Tomatoes	Cassava
1992	50	20	Tomatoes	Cassava
1993	55	20	Tomatoes	Cassava
1994	56	20	Tomatoes	Maize
1995	56	20	Tomatoes	Cassava

Water Problems on Farms

No. of Farms	Elevation (m)	Angle	Crop	Problem
15	11 - 17	$0 - 10^o$	Tomatoes	W1,2,3
		10° +	Tomatoes/maize	W1
12	17 - 29	$0 - 10^o$	Tomatoes	W1,2,3
		10° +	Tomatoes/maize	W1
13	29 - 55	$0 - 10^o$	Tomatoes	W1, D1
		10° +	Tomatoes/maize	D1
14	55- 83	$0 - 10^o$	Cassava/maize	D1, W1
		10° +	Maize	D1
15	83 - 113	$0 - 10^o$	Cassava/maize	D1
		10° +	Cassava/maize	D2
12	113 - 134	$0 - 10^o$	Cassava	D2
		10° +	Cassava/maize	D3

This table shows the results of a study of both desiccation and waterlogging on selected farms.

Classification parameters.

Desiccation (D1) Minor crops losses (<10 percent)

Desiccation (D2) Strong losses of tomatoes (30 percent +).

Desiccation (D3) Near total losses of tomatoes (these rarely planted), maize (20 – 70 percent losses).

Waterlogging (W1) Minor crop losses (<10 percent).

Waterlogging (W2) Strong losses of tomatoes, some maize (10 - 20 percent).

Waterlogging (W3) Losses of tomatoes and maize (20 – 40 percent losses).

APPENDIX 5. WATERLOGGING PRONE SOILS IN THE STUDY AREA

Soil Type	Texture	Main Location
Ustorthents		
Hacho Series	Sands	Valleys, flatland
Ustipsamments		
Doyum Series	Coarse Sands	Valleys
Nta Series	Loamy sands/sandy loams	Valleys
Endoaquents		
Papao Series	Silty clays/silty loam	Valleys, moist lowlands
Oyibi Series	Heavy clay	Valleys
Ofin Series	Sands/sandy loams	Valleys
Endoaquepts		
Krabo Series	Silty loam/silty clay	Valleys, moist lowlands

Source: Field Survey, Brammer (1967; Adu and Asiamah (1992) and SRI Report (1995).

Waterlogging Duration

To illustrate the problem of waterlogging in rainfed farms, the author took some measurements of waterlogging duration. The waterlogging problem is more serious in some years.

This is exacerbated when heavy irrigation is applied to cope with drought, and sudden rains raise the water table. This is not necessarily correlated with the total rainfall figure for the year in question. In some cases peasant farms near the Project's farms may be affected.

Site	Water Source	Initial Depth cm^2	Area m^3	Duration (days)
1	Rain	8	6.8	3
2	Rain	11	8.1	2.5
3	Rain	22	6.9	3
4	Rain	9.5	7.5	2
5	Rain	6	12.8	1
6	Rain	11	16.2	2
7	Rain	27	14.7	4.5
8	Rain	21	18.6	4
9	Rain	38	12	6
10	Rain	48	18	10
11	Rain	6	4	2
12	Rain	7	4.5	2
13	Rain	19	4.3	3
14	Rain	21	6	3
15	Rain	17	12	2
16	Rain	19	9	3
17	Rain	29	10	4
18	Rain	16	11	3
19	Rain	12	16	2
20	Rain	17	8	2

APPENDIX 6. SOIL CLASSIFICATION AND LOCAL FARMER KNOWLEDGE

The soils described above are known to the farmers in the study area, in terms of the following indicators:

1. *Soil texture* dark loamy soils are "better" than stony or sandy soils.
2. *Colour* - dark soils tend to be "better" (more fertile) than red or yellow brown soils.
3. *Differentials in topsoils and subsoils* - some soils have loamy topsoils and sandy infertile subsoils. Mechanised tillage churns these layers together, degrading the soil structure and fertility.
4. *Slope* - waterlogging occurs on flat land, especially on dark soils, while erosion occurs on slopes, especially on red soils.

5. *Clay content* - clayey soils when wet or flooded are impossible to clear or till: tractors have sunk into such soils up to a depth of 1 to 1.5 metres. A 200 pound man stepping into such soils might sink his feet about 30 to 45 centimetres (the author experimented with this method).
6. *Sand content* - sandy soils are readily desiccated; even temporary droughts can result in crop losses. Upper slope soils are too dry for water sensitive tomatoes.

Experimentation by the farmers, especially with sensitive crops such as tomato, led to such judgements. The variable costs of fertilisers, and crop changes due to changing market demand, could change their judgement of the suitability of a particular soil. Heavy rainfall years, where waterlogging and crop losses are common, favoured well drained soils (red and yellow sandy soils) and negatively affect black clays and loams. Dry years had the opposite effect. Thus the opinions expressed represented "average" assessments. The tables below relates this information to the soil types, and compares scientific knowledge with local farmer knowledge. The summarised comparison between peasant farmers' soil knowledge and official records do not differ markedly.

Alfisols

Soil Series	Official Description	Farmer Description
Fete Series	Steep slope soils, well drained, pale brown sandy rock fragments low fertility	Too stony and sandy. The low rainfall encourages desiccation; hence such soils are rarely farmed nowadays.
Salom Series	Slope soils, well drained, dark brown humus loamy and fiable, more fertile than fete series, probably due to metamorphic rather than sand stone parent rock	Good for farming, even on slopes. Still farmed even when desiccated (maize and cassava, not tomatoes). Hoes used on slopes. Identified by colour and friable loamy feel.
Adentan Series	Midslope soils, orange brown, sandy loams to sandy clays, low fertility.	Quite good for farming, though not prized. Can cause tomato losses due to desiccation during the dry season.

(Continued)

Soil Series	Official Description	Farmer Description
Adawso Series	Dark greyish brown slightly humus sandy, light loam. Fairly good for agriculture	Main soils of the Project area. Quite fertile, but less than Salom series
Bawjiasi Series	Dark reddish brown and sandy clay loams. Fairly good for agriculture.	Judged as a `red'soil. Moderately good for agriculture. Farmers prefer black and darker soils.
Truku Series	Sticky, blue grey clay. Not good for agriculture.	Not good for farming. Coconuts may be grown.

Source: SRI and Field Survey (1995).

Entisols

Soil Series	**Official Description**	**Farmer Description**
Hacho Series	Pale brown, sandy loams with low nutrient capacity. Judged suitable for agriculture	Classified with red soils. Not as good as the rich black loams. Utilised with fertiliser application. Sandy profile departed.
Toje Series	Reddish brown, non concretionary loams. Low fertility.	Not very good for farming. Tolerable but other soils prefered.
Korle Series	Red light clays. Moderately fertile.	Better than the sandy soils. However, only moderately good. Dries easily.
Keta Series	Dark grey to pale brown, loose sands. Not suitable for farming suitable for coconuts.	Not farmed. Only coconuts grown. Not fertile.
Doyum series	Pale yellow brown, course sands. Usually on steep slopes. Fertility.	Considered among less suitable soils due to sandy texture, drainage and desiccation, and general slope location

Source: SRI (1995) and Field Survey (1995 - 1996).

Soil Series	Official Description	Farmer Description
Nta Series	Very dark greyish brown, slightly humus, single grained loamy sand. Easily distinguished yellow sandy subsoil.	Common Project soil. Moderately fertile topsoil. However, tillage mixes topsoil with subsoil, decreasing fertility
Papao Series	Dark grey brown humus porous silty loam, poorly drained, subject to waterlogging. Hard when dry.	Another dark Project soil. Firly good; better than the sandy or red soils. Waterlogging a problem.
Oyibi Series	Heavy dark grey clays subjected to periodic flooding by fresh or salty water. Too saline for farming.	Generally too salty for farming. Mostly coconuts are grown.
Muni Series	Saline, alluvial lagoon clays, with grey or yellow grey sandy clay. Not suitable for farming.	Generally saline. Mostly coconuts are grown.
Ofin Series	Dark greys sands to loams. Prone to waterlogging. Quite suitable for farming.	Another common soil of the Project area. Recognised as fairly good especially in clayey areas. Waterlogging common.

Source: SRI (1995) and Field Survey.

Ultisols and Inceptisols

Soil Series	Official Description	Farmer Description
Bediesi Series (ultisol)	Dark brown to brown sandy loams. Suitable for agriculture.	Considered average for agriculture ironstone affects ploughed soils.
Mamfe Series (ultisol)	Dark brown or greyish brown humus sandy loam or clay loam. Suitable for agriculture.	Considered average for agriculture. The occurence of ironstone affects the plough soil.
Krabo Series (inceptisols)	Grey brown silty clays - dark brown silty loam. Waterlogged in wet season. Suitable for agriculture.	Fairly good. Judged above average for cultivation, but local occurences of iron pan may be churned up by tractor ploughing affecting fertility negatively.

Source: SRI (1995) and Field Survey.

APPENDIX 7. ALTERNATIVES TO FARMING, AND PROBLEMS

Activity	Main issues	Males		Females	
		40+	<40	40+	<40
Farming	Desiccation, waterlogging, cost, drought, land access	78	67	80	59
Hunting	Declining catches, habitats, youth interest, livestock alternative	68	55	45	28
Fishing	Declining catches, youth interest, livestock protein alternative, hard work	40	40	46	44
Wood cutting	Tree shrub decline, competing wood imports lower prices.	74	61	72	50
Transport	High gasoline and diesel prices, few vehicle parts, bad roads	51	63	62	63
Urban	High school fees, urban market problems, competition, urban crime	66	63	68	64
Cultural	Westernisation, declining respect for local edicts, increased youth assertiveness.	60	73	67	70

INDEX

A

B

C

D

E

F

G

H

I

J

K

L

O

P

Q

R

S

T

U

V

W

Y

Z